SpringerBriefs in Applied Sciences and Technology

Computational Intelligence

Series Editor

Janusz Kacprzyk

For further volumes:
http://www.springer.com/series/10618

Wojciech Z. Chmielowski

Management of Complex Multi-reservoir Water Distribution Systems Using Advanced Control Theoretic Tools and Techniques

Springer

Wojciech Z. Chmielowski
Institute of Water Engineering and Water Management
Krakow University of Technology
Kraków
Poland

ISSN 2191-530X ISSN 2191-5318 (electronic)
ISBN 978-3-319-00238-5 ISBN 978-3-319-00239-2 (eBook)
DOI 10.1007/978-3-319-00239-2
Springer Cham Heidelberg New York Dordrecht London

Library of Congress Control Number: 2013939065

Printed on acid-free paper

Springer is part of Springer Science+Business Media (www.springer.com)

Contents

Part I
Steady Boundary Conditions in the Trajectories of States for Optimal Management of Complex Multi-Reservoir Water Distribution System

Chapter 1
Introduction

The water deficit occurring on an ever larger scale necessitates efficient use of existing water supplies using a network of artificial and natural water reservoirs to supply countries' economic macro-regions with water. Many of these reservoirs are connected by rivers, canals or pipelines with water pumping stations, along with users consuming a particular water resources in a given hydrological area make economic water systems complex. A single water reservoir, as the primary object of a complex economic water system plays one of the most important roles in this area. Hence the need for a planned and safe execution of tasks, based on the relevant decisions concerning the regulation of the outflow of water from the reservoir and the proper use of its usable capacity. Controlling a complex system of reservoirs can be divided according to the type of decisions and time horizon to which they relate. The division identifies three main layers:

- planning long-term retention system of reservoirs,
- planning short-term retention system of reservoirs,
- operational control (ongoing).

In each layer, different rules apply for determining the retention system of the reservoirs. For planning long-term retention, the dominant role is played by methods based on the analysis of hydrological data and outflows over the longest possible period of time. Analyzing historical hydrological data in the catchment of reservoirs and water distribution data, with a specific approximation an argument can be made for long-term retention of the key elements identified by the following parameters:

- states of the filling system for reservoirs in specific divisions of time,
- a guarantee to meet the needs of users downstream of the reservoir in the intervals,
- maintaining over time the appropriate retention of flood prevention in the system.

The short-term retention plan is dependent on decisions resulting from the implementation of different types of training simulation, based on computer algorithms. The purpose of the algorithms is to allow observation of the effects of certain decision rules on the process of controlling the reservoir system (reservoir outflow vector), and thus make the right decisions in a given situation. It is necessary to distinguish

W. Z. Chmielowski, *Management of Complex Multi-reservoir Water Distribution Systems Using Advanced Control Theoretic Tools and Techniques*, SpringerBriefs in Computational Intelligence, DOI: 10.1007/978-3-319-00239-2_1, © The Author(s) 2013

clearly between short-term retention planning in normal operating conditions and regarding floods and droughts.

Notwithstanding the foregoing division, especially at the short-term retention planning layer, computer algorithms often contain computational modules using advanced optimization techniques. At this stage of formulation, solution and analysis of the results is certainly one of the essential building blocks for limiting the set of possible decisions concerning the hypothetical outflow from the reservoir. The decision making process regarding the outflow from the reservoir system, including solutions for specifically defined and conditioned optimization problems, can be used to take into account the obtained solutions as one of the building blocks for a dispatcher's final decision when setting short-term retention strategy planning.

The operational control layer is an area of operations which, in accordance with the guidelines and instructions of the previously described analyses and the assessment of the current hydrometeorological state of the catchment system in the reservoirs, the final selection decision can be made on the controlling the outflow vector. In this case, the distinction between normal operation mode and operation in emergency situations (floods and drought) is of paramount importance. Control of the current outflows from reservoirs during flooding (flood wave control) is very different as the specific working conditions, i.e. thevdramatically short period of time for taking and implementing decisions, dominate in determining the rules of procedure. Current control of a reservoir (system of reservoirs) during normal operation is to select a decision from the set of hypothetical decision rules in this layer and for this kind of operation. Often in the operational control layer, the selection of a decision is based on finding a compromise solution on the basis of optimization of the group of multi-criteria tasks. Following the decision, this time regardless of the mode of operation of the reservoir system, appropriate devices (controllers, actuators, positioners, etc.) will perform the closing or opening of the reservoir drainage equipment, thus turning the dispatcher's (person, group of people) decisions into specific activity (vector outflows from reservoirs), from which irreversible consequences follow, both at the time of implementation, as well as in subsequent time periods. The consequences of bad decisions are often catastrophic. Inadequate preparation especially of a single reservoir (earlier methodically justified lowering of the reservoir in the face of an approaching flood surge) is therefore likely to cause uncontrolled flood situation downstream of the reservoir, the effects of which may be difficult to predict. An unjustified reduction of the water in the reservoir based on unreliable estimates of inflow, in combination with the dispatcher's insufficient experience mistakenly assessing the hydrometeorological situation, in the case of low inflow to the reservoir in the immediate future the first step is to produce an emergency situation associated with a lack of adequate water supplies usually stored in reservoir system. This shows how great the responsibility rests with the employee with an impact on controlling the outflow of water from the reservoir system. Therefore, it is very risky to leave dispatcher alone with the problem of taking a decision based solely on experience, intuition, or routine, hoping that it will be adequate, and the results positive. On the other hand, the concept of "autopilot", i.e. completely replacing the dispatcher with even the most specialised computational algorithms, is completely unacceptable. . That is why over the years some automated decision-making in each of these layers of control have continually been proposed and introduced, which involves the adaptation of different support algorithms, leaving the possibility of input by the dispatcher in specific situations from his point of view. Taking the right decision at the right time is becoming one of the most important tasks. Posing the question of decision-making support for the outflow of a system of reservoirs is strongly conditioned by its location in one of the three aforementioned layers. This task is far more complicated with increasing the dimensionality of the problem. We are then talking about outflow vectors from reservoirs, structurally and task related within a complex system of water supply and management. The structure of the system takes into account all the possible connections between the reservoirs and water customers. What is usually analysed are variations of use of the water sources in the reservoirs during the operation and taking into account the effects of prospective management of the reservoirs caused by the current control system, according to a decision by the dispatcher recognised as the most appropriate in a given situation. The dispatcher's decisions determining control system of outflows from the reservoirs contain elements of such issues as multi-criteria, randomness, dynamics, non-linearity and the scale of complexity complementing the factors directly linked to the dispatcher (including attitudes to risk). The issue related to decision-making in water management systems is characterised by its interdisciplinary nature, and the quality control of the system in this regard can only be achieved by a comparative analysis of the system's historical hydrological data in the long term. Therefore, the need to eliminate bad dispatcher decisions is beyond question by:

- continuous, multidirectional development of methodologies and decision support algorithms for planning for all layers of reservoir retention, leading to the application of the latest achievements in science and technology in this area,
- The selection of dispatchers with the relevant experience and relevant mental and physical characteristics which in times of crisis may be a necessary precondition and guarantee of a proper conduct.

This study discusses issues of optimal water management in a distribution system. The main elements of the complex water-management system under consideration are retention reservoirs, among which water transfers are possible, and a network of connections between these reservoirs and water treatment plants (WTPs). System operation optimisation involves determining the proper water transport routes and their flow volumes from the retention reservoirs to the WTPs, and the volumes of possible transfers among the reservoirs, taking into account transport-related delays for inflows, outflows and water transfers in the system. Total system operation costs defined by an assumed quality coefficient should be minimal. An analytical solution of the optimisation task so formulated has been obtained as a result of using Pontriagin's maximum principle with reference to the quality coefficient assumed. Stable start and end conditions in reservoir state trajectories have been assumed. The researchers have taken into account cases of steady (Chap. 2) and transient (Chap. 3) optimisation duration. The solutions obtained have enabled the creation of computer models simulating system operation. In future, an analysis of the results obtained may affect decisions supporting the control of currently existing complex water-management systems.

Chapter 3 is a continuation of the study presented in Chap. 2 and forms its continuation, this time referring to a transient duration for optimising the operation of a system of retention reservoirs. Two essential cases in this regard are discussed—the search for the optimal time to commence reservoir system operation with a fixed end time, and the search for the optimal time to end the reservoir operation with a fixed time for system operation start. Analysis of both cases with reference to a system of cooperating dynamic facilities, such as water reservoirs, has shown that in consequence the search for optimisation commencement times is determined by a vector with optimal elements and, usually, different operation commencement times for successive reservoirs as part of the whole system operation. By analogy, while optimising operation end time for a reservoir system, a vector has been determined whose elements are operation end times for individual reservoirs in the system which usually different owing to the assumed parameters of reservoirs and input data for the system operation optimisation.

The first step towards recording and analysing the control of water-management system with a generalised structure of connections among the system elements will involve tracing the solution for water distribution from a system of reservoirs with an interlinked structure and a specific number of elements, both with reference to reservoirs and to system water consumers (Water Treatment Plants), taking into account the following conditions in trajectories of reservoir states and with reference to the optimisation horizon,

- SLC—steady left conditions in state trajectories,
- SRC—steady right conditions in state trajectories,
- ST—steady optimisation time,
- TT—transient (free) optimisation time.

As regards the last case, there are two further issues:

- Free Start Time (FST), for steady optimisation end time,
- Free End Time (FET), for steady optimisation start.

Chapter 2
Steady Optimisation Time, ST

2.1 Steady Optimisation Time, ST

Figure 2.1 presents a model system of three interconnected reservoirs supplying water to independent consumers. The end conditions in the reservoir state trajectories are steady. This means that after an optimisation period W, the reservoir states should match previously determined values. For the vector of predicted inflows to the reservoirs, the optimising task formulated with index (2.1) comes down to:

- defining the control vector (outflows from reservoirs),

$$\hat{\boldsymbol{u}}(t - \boldsymbol{h}(t))_{(+)}, \quad \forall t \in [t_0, W] \tag{2.1}$$

  which will minimally differ from the vector consisting of partial water demands per individual system reservoir,

$$\mathbf{B}(t) \cdot \mathbf{Y}(t) \cdot \mathbf{S} \cdot \boldsymbol{1}, \forall t \in [t_0, W] \tag{2.2}$$

- achieving the target specified above at minimum transfer cost among reservoirs,
- obtaining at the end of optimisation horizon W the vector for the filling levels for the reservoirs $\boldsymbol{x}(W)$ satisfying the required values,

The reservoirs supply water to four consumers (WTPs) at the same time; therefore, to describe the system any further it is necessary to introduce a function for reservoir involvement in carrying out water demand functions $Y_j(t),\ j = 1, \ldots, 4\ \forall t \in [t_0, W]$ dividing (for each instant in time) the function $Y_j(t),\ j = 1, \ldots, 4$ among the reservoirs in the system:

$$Y_j(t) - \sum_{i=1}^{4} b_{i,j}(t) \cdot Y_j = 0, \quad j = 1, \ldots, 4 \tag{2.3}$$

W. Z. Chmielowski, *Management of Complex Multi-reservoir Water Distribution Systems Using Advanced Control Theoretic Tools and Techniques*, SpringerBriefs in Computational Intelligence, DOI: 10.1007/978-3-319-00239-2_2, © The Author(s) 2013

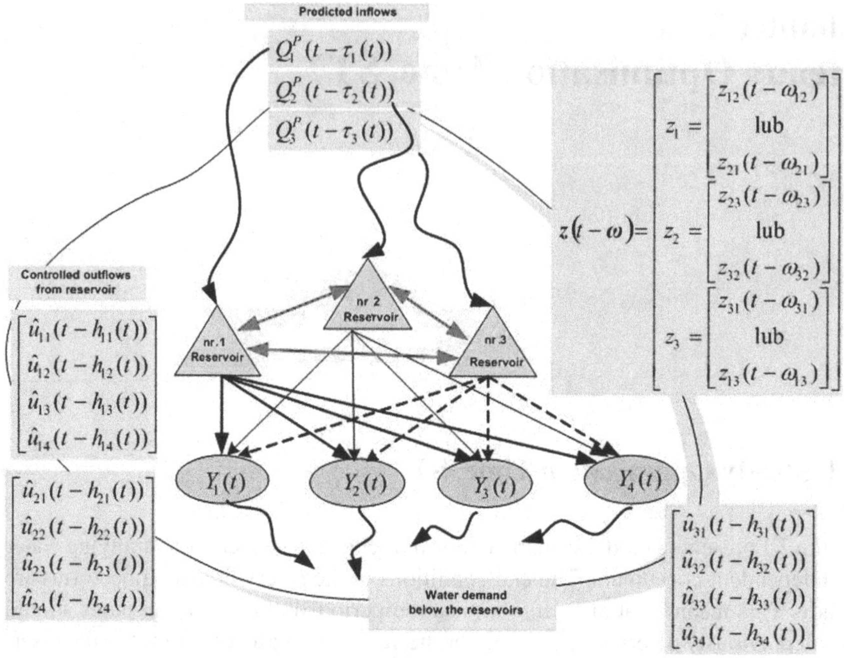

Fig. 2.1 The complex water-management system

In this problem of reservoir system functioning (Fig. 2.1), there are a number of delays in individual system input and output variables. The abovementioned delays, which substantially complicate the formal record of system function, come down to the following dependencies:

- vector of delays related to water flow through the reservoirs

$$\tau(t)^{\mathrm{T}} = [\tau_1(t)\ \ \tau_2(t)\ \ \tau_3(t)] \tag{2.4}$$

The water flow rate measurement point $[\mathrm{m}^3/\mathrm{s}]$ is usually located in the area where a river flows into a reservoir. Flow rates measured at the reservoir inlet will appear in the vicinity of a dam after a given time, that is, with a delay dependent on the reservoir's dimensions. The delay is also a function of time, because it may change with hydrological and climatic conditions within the reservoir area (higher, lower, variable rate of water flow through reservoir).
Therefore, the vector of inflows into the system reservoirs taking into account the abovementioned delays should be defined as follows:

$$\boldsymbol{Q}^P(t-\tau(t)) = \begin{bmatrix} Q_1^P(t-\tau_1(t)) \\ Q_2^P(t-\tau_2(t)) \\ Q_3^P(t-\tau_3(t)) \end{bmatrix} \tag{2.5}$$

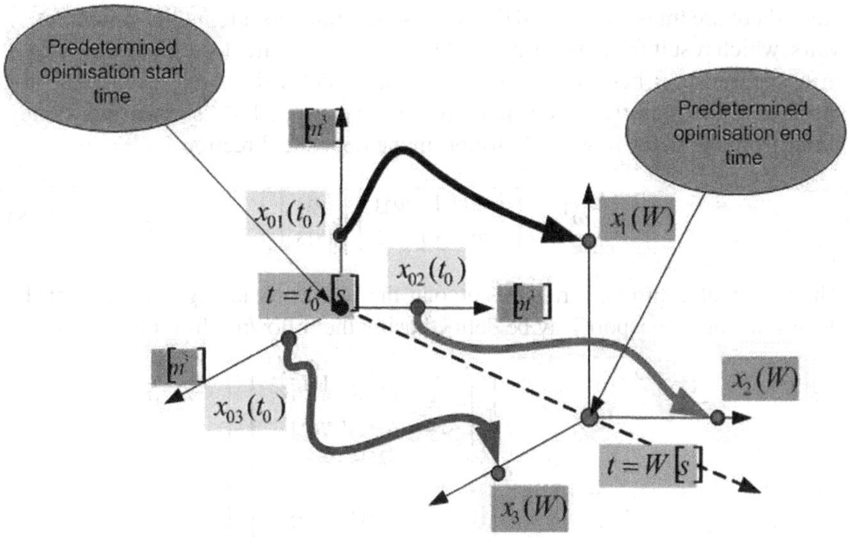

Fig. 2.2 Optimisation start and end time

- Another are transport-related delays in water distribution from the reservoir system to the WTPs (2.6). These delays result from the distance between individual reservoirs and water consumers. The vector of controlled outflows from reservoirs taking into account the abovementioned delays is defined by the following formula (2.7):

$$\boldsymbol{h}(t) = \begin{bmatrix} h_{11}(t) & h_{21}(t) & h_{31}(t) \\ h_{12}(t) & h_{22}(t) & h_{32}(t) \\ h_{13}(t) & h_{23}(t) & h_{33}(t) \\ h_{14}(t) & h_{24}(t) & h_{34}(t) \end{bmatrix} \tag{2.6}$$

$$\hat{\boldsymbol{u}}(t - \boldsymbol{h}(t))_{(+)} = \begin{bmatrix} \begin{bmatrix} u_{11}(t - h_{11}(t)) \\ u_{12}(t - h_{12}(t)) \\ u_{13}(t - h_{13}(t)) \\ u_{14}(t - h_{14}(t)) \end{bmatrix} \\ \begin{bmatrix} u_{21}(t - h_{21}(t)) \\ u_{22}(t - h_{22}(t)) \\ u_{23}(t - h_{23}(t)) \\ u_{24}(t - h_{24}(t)) \end{bmatrix} \\ \begin{bmatrix} u_{31}(t - h_{31}(t)) \\ u_{32}(t - h_{32}(t)) \\ u_{33}(t - h_{33}(t)) \\ u_{34}(t - h_{34}(t)) \end{bmatrix} \end{bmatrix} \tag{2.7}$$

- Also, there are transport-related delays in water transfers among the system reservoirs, which result from the location of reservoirs in the field. Additionally, delays in water transport between reservoirs are asymmetrical, which means that, for instance, a delay in water transport from reservoir no. 1 to reservoir no. 2 may differ from the delay in water transport in the opposite direction.

$$\boldsymbol{\omega}^{\mathrm{T}} = \left[\begin{bmatrix}\omega_{12}\\ \omega_{21}\end{bmatrix}\begin{bmatrix}\omega_{23}\\ \omega_{32}\end{bmatrix}\begin{bmatrix}\omega_{31}\\ \omega_{13}\end{bmatrix}\right] \tag{2.8}$$

The vector of controlled transfers among the reservoirs taking into account the delays in water transport may be defined using the following formula:

$$\hat{z}(t-\boldsymbol{\omega}) = \begin{bmatrix} \begin{bmatrix} z_1 = & \begin{matrix} z_{12}(t-\omega_{12}) \\ or \\ z_{21}(t-\omega_{21}) \end{matrix} \end{bmatrix} \\ \\ \begin{bmatrix} z_2 = & \begin{matrix} z_{23}(t-\omega_{23}) \\ or \\ z_{32}(t-\omega_{32}) \end{matrix} \end{bmatrix} \\ \\ \begin{bmatrix} z_3 = & \begin{matrix} z_{31}(t-\omega_{31}) \\ or \\ z_{13}(t-\omega_{13}) \end{matrix} \end{bmatrix} \end{bmatrix} \tag{2.9}$$

2.1.1 Quality Coefficient

Optimal reservoir state trajectories and trajectories of outflows from reservoirs (control room) will be determined through solving the dynamic optimising task specified below. Its requisite elements include:

$$F = 0,5 \int_{t_0}^{W} \left\{ \begin{matrix} [\mathbf{B}(t)\cdot\mathbf{Y}(t)\cdot\mathbf{S}\cdot\mathbf{1} - \boldsymbol{u}(t-\boldsymbol{h}(t))]^{\mathrm{T}}_{(+)} \\ \cdot\mathbf{A}_1\cdot \\ [\mathbf{B}(t)\cdot\mathbf{Y}(t)\cdot\mathbf{S}\cdot\mathbf{1} - \boldsymbol{u}(t-\boldsymbol{h}(t))]_{(+)} \\ \cdot\mathbf{z}(t-\boldsymbol{\omega})^{\mathrm{T}}\cdot\mathbf{A}_2\cdot\mathbf{z}(t-\boldsymbol{\omega}) \end{matrix} \right\} dt \tag{2.10}$$

Regarding Fig. 2.1, the following symbols are used in Eq. (2.10):

- Matrix $\mathbf{B(t)}$ is a diagonal block matrix with terms constituting diagonal matrixes. Their elements are functions of reservoir involvement $i = 1, \ldots, 3$ in carrying out the demand functions $Y_j(t)$, $j = 1, \ldots 4$, $[m^3/s]$.

$$\mathbf{B(t)} = \begin{bmatrix} \mathbf{B}_1(t) & * & * \\ * & \mathbf{B}_2(t) & * \\ * & * & \mathbf{B}_3(t) \end{bmatrix} \tag{2.11}$$

where:

$$\mathbf{B}_1(t) = \begin{bmatrix} b_{11}(t) & 0 & 0 & 0 \\ 0 & b_{12}(t) & 0 & 0 \\ 0 & 0 & b_{13}(t) & 0 \\ 0 & 0 & 0 & b_{14}(t) \end{bmatrix} \quad (2.12)$$

$$\mathbf{B}_2(t) = \begin{bmatrix} b_{21}(t) & 0 & 0 & 0 \\ 0 & b_{22}(t) & 0 & 0 \\ 0 & 0 & b_{23}(t) & 0 \\ 0 & 0 & 0 & b_{24}(t) \end{bmatrix} \quad (2.13)$$

$$\mathbf{B}_3(t) = \begin{bmatrix} b_{31}(t) & 0 & 0 & 0 \\ 0 & b_{32}(t) & 0 & 0 \\ 0 & 0 & b_{33}(t) & 0 \\ 0 & 0 & 0 & b_{34}(t) \end{bmatrix} \quad (2.14)$$

$$* = \begin{bmatrix} 0\,0\,0\,0 \\ 0\,0\,0\,0 \\ 0\,0\,0\,0 \\ 0\,0\,0\,0 \end{bmatrix} \quad (2.15)$$

Matrix $\mathbf{Y}(t)$ is a diagonal block matrix with terms constituting diagonal matrixes. Their elements are the demand functions effective in the system $Y_j(t)$, $j = 1, \ldots 4$, $[m^3/s]$ Fig. 2.2

$$\mathbf{Y}(t) = \begin{bmatrix} \circ(t) & * & * \\ * & \circ(t) & * \\ * & * & \circ(t) \end{bmatrix} \quad (2.16)$$

where:

$$\circ(t) = \begin{bmatrix} Y_1(t) & 0 & 0 & 0 \\ 0 & Y_2(t) & 0 & 0 \\ 0 & 0 & Y_3(t) & 0 \\ 0 & 0 & 0 & Y_4(t) \end{bmatrix} \quad (2.17)$$

- The control vector (of controlled outflows from reservoirs) (7) is a block vector. Its elements are vectors, and their elements are reservoir outflows $i = 1, \ldots, 3$ to conurbations $j = 1, \ldots, 4$.
- Then, matrix $\mathbf{A_1}$ is a positively definite block matrix with terms on a diagonal also being diagonal matrixes. Their elements are weight coefficients related to proper control vector elements

$$\mathbf{A_1} = \begin{bmatrix} \bullet_1 & * & * \\ * & \bullet_2 & * \\ * & * & \bullet_3 \end{bmatrix} \quad (2.18)$$

where:

$$\bullet_1 = \begin{bmatrix} a_{11} & 0 & 0 & 0 \\ 0 & a_{12} & 0 & 0 \\ 0 & 0 & a_{13} & 0 \\ 0 & 0 & 0 & a_{14} \end{bmatrix} \tag{2.19}$$

$$\bullet_2 = \begin{bmatrix} a_{21} & 0 & 0 & 0 \\ 0 & a_{22} & 0 & 0 \\ 0 & 0 & a_{23} & 0 \\ 0 & 0 & 0 & a_{24} \end{bmatrix} \tag{2.20}$$

$$\bullet_3 = \begin{bmatrix} a_{31} & 0 & 0 & 0 \\ 0 & a_{32} & 0 & 0 \\ 0 & 0 & a_{33} & 0 \\ 0 & 0 & 0 & a_{34} \end{bmatrix} \tag{2.21}$$

- Further, positively definite diagonal matrix $\mathbf{A_2}$,

$$\mathbf{A_2} = \begin{bmatrix} a_{11} & 0 & 0 \\ 0 & a_{22} & 0 \\ 0 & 0 & a_{33} \end{bmatrix} \tag{2.22}$$

with elements, which are weight coefficients related to a subintegral section of the quality coefficient (2,10), corresponding to the costs of water transfers among the reservoirs, e.g. a_{11} all applies to transfer cost $z_1(t)$, that is from reservoir 1 to reservoir 2, or the other way round, etc. Regarding the matrix, this way of presenting the problem is a simplification, because in a general case the water transfer cost e.g. from reservoir no. 1 to reservoir no. 2 will not necessarily equal the cost of water transfer from reservoir no. 2 to reservoir no. 1 (e.g. gravitational flow and pumping: $a_{11}^{zb1 \rightarrow zb2} \neq a_{11}^{zb1 \leftarrow zb2}$).
- Then, unit vector

$$\boldsymbol{I}^{\mathrm{T}} = \left[\otimes \otimes \otimes \right], \quad \otimes = \left[1\ 1\ 1\ 1 \right] \tag{2.23}$$

- and vector of transfers among reservoirs (2.9)

2.1.2 State Equation of Reservoirs

The following have been assumed: balance equation of state for the system reservoirs, and start and end conditions in reservoir state trajectories

$$f : \dot{\boldsymbol{x}}(t) = \boldsymbol{Q}^P(t - \boldsymbol{\tau}(t)) - \mathbf{S_1}\boldsymbol{u}(t - \boldsymbol{h}(t)) + \mathbf{S_2}\boldsymbol{z}(t - \boldsymbol{\omega}) \tag{2.24}$$

$$x_0(t_0)^{\mathrm{T}} = \left[x_{01}(t_0)\ x_{02}(t_0)\ x_{03}(t_0) \right] \tag{2.25}$$

$$x^U(W)^{\mathrm{T}} = \left[x_1{}^U(W)\ x_2{}^U(W)\ x_3{}^U(W) \right] \tag{2.26}$$

The following symbols are used in state equation (2.24):

- vector of reservoir state derivatives

$$\dot{x}^{\mathrm{T}}(t) = [dx_1(t)/dt\ \ dx_2(t)/dt\ \ dx_3(t)/dt] \tag{2.27}$$

- vector of predicted inflows to reservoirs (2.5) $\left[\mathrm{m}^3/\mathrm{s}\right]$
- vector of predetermined initial filling levels in reservoirs (2.25) $\left[\mathrm{m}^{\mathrm{s}}\right]$
- vector of predetermined final filling levels in reservoirs (2.26) $\left[\mathrm{m}^{\mathrm{s}}\right]$
- t_0 predetermined optimisation start time $[s]$
- W predetermined optimisation end time $[s]$
- diagonal structural matrix $\mathbf{S}_1$, needed to record the system structure and the system state equation

$$\mathbf{S}_1 = \begin{bmatrix} \otimes & [0] & [0] \\ [0] & \otimes & [0] \\ [0] & [0] & \otimes \end{bmatrix} \quad [0] = \left[0\,0\,0\,0\right] \tag{2.28}$$

- structural matrix needed to record the relations between reservoirs with reference to transfers among reservoirs in the system state equation

$$\mathbf{S}_2 = \begin{bmatrix} -1 & 0 & 1 \\ 1 & -1 & 0 \\ 0 & 1 & -1 \end{bmatrix} \tag{2.29}$$

- Matrix $\mathbf{S}$ is a structural matrix derived from operation $\mathbf{S} = ((\mathbf{S}_1{}^{\mathrm{T}} \cdot \mathbf{S}_1) * \mathbf{I})$, (asterisk $*$, table multiplication).

2.1.3 Solution for the Optimising Task

Hamilton's function for the system of Eqs. (2.10) and (2.24) has the following form:

$$H = -f_0 + \boldsymbol{\psi}^{\mathrm{T}} \cdot \boldsymbol{f} \tag{2.30}$$

f_0 —sub-integral function of coefficient (2.10)
f —state equation for reservoirs (2.24)
$\boldsymbol{\psi}$ —conjugate variable, vector [3*1]

$$H = -0,5 \cdot \left\{ \begin{array}{l} [\mathbf{B}(t) \cdot \mathbf{Y}(t) \cdot \mathbf{S} \cdot \mathbf{1} - \boldsymbol{u}(t - \boldsymbol{h}(t))]_{(+)}^{\mathrm{T}} \\ \cdot \mathbf{A}_1 \cdot \\ [\mathbf{B}(t) \cdot \mathbf{Y}(t) \cdot \mathbf{S} \cdot \mathbf{1} - \boldsymbol{u}(t - \boldsymbol{h}(t))]_{(+)} \\ \cdot \mathbf{z}(t-\omega)^{\mathrm{T}} \cdot \mathbf{A}_2 \cdot \mathbf{z}(t-\omega) \end{array} \right\} + \boldsymbol{\psi}(t)^{\mathrm{T}} \cdot \left[\boldsymbol{Q}^P(t - \boldsymbol{\tau}(t)) - \mathbf{S}_1 \boldsymbol{u}(t - \boldsymbol{h}(t)) + \mathbf{S}_2 \boldsymbol{z}(t-\omega) \right] \quad (2.31)$$

The system of equations for Hamilton's function in form (2.31) is shown below:

1.

$$\left[(\nabla_u H)_{\hat{u},\hat{x},\hat{\psi}} \right]^{\mathrm{T}} = \boldsymbol{0} \Rightarrow \hat{\boldsymbol{u}}(t - \boldsymbol{h}(t))_{(+)} = \mathbf{B}(t) \cdot \mathbf{Y}(t) \cdot \mathbf{S} \cdot \mathbf{1} - \mathbf{A}_1^{-1} \cdot \mathbf{S}_1{}^{\mathrm{T}} \cdot \hat{\boldsymbol{\psi}}(t) \quad (2.32)$$

2.

$$\left[(\nabla_z H)_{\hat{u},\hat{z},\hat{x},\hat{\psi}} \right]^{\mathrm{T}} = \boldsymbol{0} \Rightarrow \hat{\mathbf{z}}(t-\omega) = \mathbf{A}_2^{-1} \cdot \mathbf{S}_2^{\mathrm{T}} \cdot \hat{\boldsymbol{\psi}}(t) \quad (2.33)$$

3.

$$\left[\left(\nabla_\psi H \right)_{\hat{u},\hat{x}} \right]^{\mathrm{T}} = \dot{\boldsymbol{x}}(t) \Rightarrow \boldsymbol{Q}^P(t - \boldsymbol{\tau}(t)) - \mathbf{S}_1 \boldsymbol{u}(t - \boldsymbol{h}(t)) + \mathbf{S}_2 \boldsymbol{z}(t-\omega) \quad (2.34)$$

4.

$$\left[-(\nabla_x H)_{\hat{u},\hat{x},\hat{\psi}} \right]^{\mathrm{T}} = \dot{\hat{\boldsymbol{\psi}}}(t) \Rightarrow \dot{\hat{\boldsymbol{\psi}}}(t) = \boldsymbol{0}_{(3*1)} \quad (2.35)$$

Equation (2.35) proves that:

$$\hat{\boldsymbol{\psi}}(t) = \boldsymbol{C}_1 \quad (2.36)$$

Integration of Eq. (2.34) enables us to obtain an equation characterising the general form of a state trajectory vector:

$$\hat{\boldsymbol{x}}(t) = \int_{t_0}^{t} \left\{ \begin{array}{l} \boldsymbol{Q}^P(\xi - \boldsymbol{h}(\xi)) + \\ -\mathbf{S}_1 \cdot \hat{\boldsymbol{u}}(\xi - \boldsymbol{h}(\xi))_{(+)} \\ +\mathbf{S}_2 \cdot \hat{z}(\xi - \omega) \end{array} \right\} d\xi + \boldsymbol{C}_2 \quad (2.37)$$

We substitute term (2.36) in Eq. (2.32), and then the control vector is defined by the following equation:

$$\hat{\boldsymbol{u}}(t - \boldsymbol{h}(t))_{(+)} = \mathbf{B}(t) \cdot \mathbf{Y}(t) \cdot \mathbf{S} \cdot \mathbf{1} - \mathbf{A}_1^{-1} \cdot \mathbf{S}_1^{\mathrm{T}} \cdot \boldsymbol{C}_1 \quad (2.38)$$

We substitute term (2.36) in Eq. (2.33), and then the vector of controlled transfers among reservoirs is defined by the following equation:

$$\hat{\mathbf{z}}(t-\omega) = \mathbf{A}_2^{-1} \cdot \mathbf{S}_2^{\mathrm{T}} \cdot \boldsymbol{C}_1 \tag{2.39}$$

Then, we attempt to determine the state vector. We substitute (2.38) and (2.39) in (2.37), thus receiving

$$\hat{\boldsymbol{x}}(t) = \int\limits_{t_0}^{t} \left\{ \begin{array}{l} \boldsymbol{Q}^P(\xi - \boldsymbol{h}(\xi)) \\ -\mathbf{S}_1 \cdot [\mathbf{B}(\xi) \cdot \mathbf{Y}(\xi) \cdot \mathbf{S} \cdot \boldsymbol{1}] \\ +\mathbf{S}_1 \cdot \left[\mathbf{A}_1^{-1} \cdot \mathbf{S}_1^{\mathrm{T}} \cdot \boldsymbol{C}_1\right] \\ +\mathbf{S}_2 \cdot \left[\mathbf{A}_2^{-1} \cdot \mathbf{S}_2^{\mathrm{T}} \cdot \boldsymbol{C}_1\right] \end{array} \right\} d\xi + \boldsymbol{C}_2 \tag{2.40}$$

for $t = t_0 \Rightarrow \boldsymbol{C}_2 = \boldsymbol{x}(t_0)$.

After regrouping and integrating the component constant in time, and for $t = W$, the state trajectory vector assumes the following form:

$$\boldsymbol{x}(W) = \int\limits_{t_0}^{W} \left\{ \begin{array}{l} \boldsymbol{Q}^P(t - \boldsymbol{h}(t)) \\ -\mathbf{S}_1 \cdot [\mathbf{B}(t) \cdot \mathbf{Y}(t) \cdot \mathbf{S} \cdot \boldsymbol{1}] \end{array} \right\} dt$$
$$+ \left(\mathbf{S}_1 \cdot \mathbf{A}_1^{-1} \cdot {\mathbf{S}_1}^{\mathrm{T}} + \mathbf{S}_2 \cdot \mathbf{A}_2^{-1} \cdot \mathbf{S}_2^{\mathrm{T}}\right) \cdot \boldsymbol{C}_1 \cdot (W - t_0) + \boldsymbol{x}(t_0) \tag{2.41}$$

We calculate the vector of constants $\boldsymbol{C}_1$ from Eq. 2.41)

$$\boldsymbol{C}_1 = \left[\left(\mathbf{S}_1 \cdot \mathbf{A}_1^{-1} \cdot \mathbf{S}_1^{\mathrm{T}} + \mathbf{S}_2 \cdot \mathbf{A}_2^{-1} \cdot \mathbf{S}_2^{\mathrm{T}}\right) \cdot (W - t_0)\right]^{-1} \cdot \left\{ \begin{array}{l} \boldsymbol{x}(W) - \boldsymbol{x}(t_0) - \\ + \int\limits_{t_0}^{W} \left[\boldsymbol{Q}^P(t - \boldsymbol{h}(t)) - \mathbf{S}_1 \cdot \mathbf{B}(t) \cdot \mathbf{Y}(t) \cdot \mathbf{S} \cdot \boldsymbol{1}\right] dt \end{array} \right\} \tag{2.42}$$

Further solving of the problem does not represent any major difficulties—substitute (2.42) in (2.36), and then obtained result in (2.32) and (2.33), thus receiving optimal control vector $\hat{\boldsymbol{u}}(t - \boldsymbol{h}(t))_{(+)}$, $\forall t \in [t_0, W]$, and vector of transfers among reservoirs $\hat{\mathbf{z}}(t - \omega)$. Transport-related delays in transfers among reservoirs essentially affect satisfying end condition of reservoir state trajectories (2.26). They force changes in transfer vector value so as to fulfil condition (2.26). Satisfying prerequisite condition (2.26) involves a need to introduce modified water distribution among the reservoirs. This change is a consequence of the scheme shown below. Formula (2.43) determines adequately increased/reduced transfer among the reservoirs during the distribution period, taking into account the delays.

$$\mathbf{z}(t-\omega) = \begin{bmatrix} z_1(t - \omega_{12 \vee 21}) : \otimes_1 \\ z_2(t - \omega_{23 \vee 32}) : \otimes_2 \\ z_3(t - \omega_{31 \vee 13}) : \otimes_3 \end{bmatrix}$$

$$\otimes_1 = \left\{ \begin{array}{l} |z| = |z_{12}| = |z_{21}| \\ (\text{sgn}(z_1) = 1 \ \wedge \ \omega_{12} > 0) \ \Rightarrow \\ z_1(t - \omega_{12}) = |z_{12}| + \frac{\int_0^{\omega_{12}} |z_{12}| dt}{(W - \omega_{12})}; \\ (\text{sgn}(z_1) = -1 \ \wedge \ \omega_{21} > 0) \ \Rightarrow \\ z_1(t - \omega_{21}) = |z_{12}| + \frac{\int_0^{\omega_{21}} |z_{12}| dt}{(W - \omega_{21})} \end{array} \right\} \tag{2.43}$$

$$\otimes_2 = \left\{ \begin{array}{l} |z| = |z_{23}| = |z_{32}| \\ (\text{sgn}(z_2) = 1 \ \wedge \ \omega_{23} > 0) \ \Rightarrow \\ z_2(t - \omega_{23}) = |z_{23}| + \frac{\int_0^{\omega_{23}} |z_{23}| dt}{(W - \omega_{23})} \\ (\text{sgn}(z_2) = -1 \ \wedge \ \omega_{32} > 0) \ \Rightarrow \\ z_2(t - \omega_{32}) = |z_{32}| + \frac{\int_0^{\omega_{32}} |z_{23}| dt}{(W - \omega_{32})} \end{array} \right\} \tag{2.44}$$

$$\otimes_3 = \left\{ \begin{array}{l} |z| = |z_{31}| = |z_{13}| \\ (\text{sgn}(z_3) = 1 \ \wedge \ \omega_{31} > 0) \ \Rightarrow \\ z_3(t - \omega_{31}) = |z_{31}| + \frac{\int_0^{\omega_{31}} |z_{31}| dt}{(W - \omega_{31})} \\ (\text{sgn}(z_3) = -1 \ \wedge \ \omega_{13} > 0) \ \Rightarrow \\ z_3(t - \omega_{13}) = |z_{13}| + \frac{\int_0^{\omega_{13}} |z_{13}| dt}{(W - \omega_{13})} \end{array} \right\}$$

Formulas (2.43, 2.44) take into account the possibility of water transfer both ways (e.g. from reservoir no. 1 to reservoir no. 2 and vice versa) and asymmetrical transport-related delay (np. $\omega_{12} \neq \omega_{21}$). This principle is applicable to all relations among the reservoirs. As regards this water system (Fig. 2.1), notation of transfer vector modifications $\hat{\mathbf{z}}(t - \boldsymbol{\omega})$ formula (2.43, 2.44), is relatively simple. Its complexity degree increases very quickly with the system dimensionality, and is particularly dependent on the number and direction of transfers among reservoirs and the delay involved.

Having completed the abovementioned modifications, we substitute (2.38) and (2.39) in (2.41) to obtain the vector of optimal state trajectories satisfying condition (2.30). We receive the minimum quality coefficient value by substituting (2.38) and (2.30) in (2.10).

2.1.4 Computer Simulations

Hipothetical scenario of events. Example 1 In order to test whether the solution obtained is optimal and correct, in the first example we will make the following simple numerical assumptions, specifying that there will be no delays in inflow and further distribution of water:

- initial/start states in reservoirs

$$\boldsymbol{x}_0(t_0)^{\mathrm{T}} = \left[\,10\ 10\ 10\,\right]\left[m^3\right]$$

- final/end states in reservoirs $\boldsymbol{x}^U(W)^{\mathrm{T}} = \left[\,30\ 30\ 30\,\right]\left[m^3\right]$
- inflows to reservoirs

$$\boldsymbol{Q}^P(t-\boldsymbol{\tau}(t)) = \begin{bmatrix} (t-(\tau_1=0)+1) \\ (t-(\tau_2=0)+1) \\ (t-(\tau_3=0)+1) \end{bmatrix}\left[m^3/s\right]$$

- water demand below reservoirs

$$\mathbf{Y}(t) = \begin{bmatrix} [*] & 0^* & 0^* \\ 0^* & [*] & 0^* \\ 0^* & 0^* & [*] \end{bmatrix}\left[m^3/s\right]$$

$$[*] = \begin{bmatrix} 4(t) & 0 & 0 & 0 \\ 0 & 4(t) & 0 & 0 \\ 0 & 0 & 4(t) & 0 \\ 0 & 0 & 0 & 4(t) \end{bmatrix}$$

$$0^* = \begin{bmatrix} 0\,0\,0\,0 \\ 0\,0\,0\,0 \\ 0\,0\,0\,0 \\ 0\,0\,0\,0 \end{bmatrix}$$

- the functions of reservoir involvement in carrying out the demand functions

$$\mathbf{B}(t) = \begin{bmatrix} [*_1] & 0^* & 0^* \\ 0^* & [*_1] & 0^* \\ 0^* & 0^* & [*_1] \end{bmatrix}$$

$$[*_1] = \begin{bmatrix} 0.33 & 0 & 0 & 0 \\ 0 & 0.33 & 0 & 0 \\ 0 & 0 & 0.33 & 0 \\ 0 & 0 & 0 & 0.33 \end{bmatrix}\cdot(t)$$

- matrices of weight coefficients

$$\mathbf{A}_1 = \begin{bmatrix} [*] & 0^* & 0^* \\ 0^* & [*] & 0^* \\ 0^* & 0^* & [*] \end{bmatrix}, \quad [*] = \begin{bmatrix} 1\,0\,0\,0 \\ 0\,1\,0\,0 \\ 0\,0\,1\,0 \\ 0\,0\,0\,1 \end{bmatrix}$$

$$\mathbf{A}_2 = \begin{bmatrix} 1\,0\,0 \\ 0\,1\,0 \\ 0\,0\,1 \end{bmatrix}$$

- structural matrix

$$\mathbf{S}_1 = \begin{bmatrix} [*] & 0^{\bullet} & 0^{\bullet} \\ 0^{\bullet} & [*] & 0^{\bullet} \\ 0^{\bullet} & 0^{\bullet} & [*] \end{bmatrix}$$

$$[*] = \begin{bmatrix} 1\,1\,1\,1 \end{bmatrix}, \quad 0^{\bullet} = \begin{bmatrix} 0\,0\,0\,0 \end{bmatrix}$$

- structural matrix

$$\mathbf{S}_2 = \begin{bmatrix} -1 & 0 & 1 \\ 1 & -1 & 0 \\ 0 & 1 & -1 \end{bmatrix}$$

- start time $t_0 = 0[s]$, same for all reservoirs,
- end time $W = 10[s]$, same for all reservoirs,

Solution

- acc. to (2.42) $\boldsymbol{C}_1 = [0,32 \quad 0,32 \quad 0,32]$
- acc. to (2.38) $\hat{\boldsymbol{u}}(t - \boldsymbol{h}(t))_{(+)} = \mathbf{B}(t) \cdot \mathbf{Y}(t) \cdot \mathbf{S} \cdot \boldsymbol{1} - \mathbf{A}_1^{-1} \cdot {\mathbf{S}_1}^{\mathrm{T}} \cdot \boldsymbol{C}_1$

$$\hat{\boldsymbol{u}}(t - \boldsymbol{h}(t))_{(+)} = \begin{bmatrix} \begin{bmatrix} 1.32 \\ 1.32 \\ 1.32 \\ 1.32 \end{bmatrix} \\ \begin{bmatrix} 1.32 \\ 1.32 \\ 1.32 \\ 1.32 \end{bmatrix} \\ \begin{bmatrix} 1.32 \\ 1.32 \\ 1.32 \\ 1.32 \end{bmatrix} \end{bmatrix} - \begin{bmatrix} \begin{bmatrix} 0,32 \\ 0,32 \\ 0,32 \\ 0,32 \end{bmatrix} \\ \begin{bmatrix} 0,32 \\ 0,32 \\ 0,32 \\ 0,32 \end{bmatrix} \\ \begin{bmatrix} 0,32 \\ 0,32 \\ 0,32 \\ 0,32 \end{bmatrix} \end{bmatrix} = \begin{bmatrix} \begin{bmatrix} 1.0 \\ 1.0 \\ 1.0 \\ 1.0 \end{bmatrix} \\ \begin{bmatrix} 1.0 \\ 1.0 \\ 1.0 \\ 1.0 \end{bmatrix} \\ \begin{bmatrix} 1.0 \\ 1.0 \\ 1.0 \\ 1.0 \end{bmatrix} \end{bmatrix}$$

- acc. to (2.39) $\hat{z}(t - \omega) = \mathbf{A}_2^{-1} \cdot \mathbf{S}_2^{\mathrm{T}} \cdot \boldsymbol{C}_1$

$$\hat{\mathbf{z}}(t - \omega) = \begin{bmatrix} 1\,0\,0 \\ 0\,1\,0 \\ 0\,0\,1 \end{bmatrix} \cdot \begin{bmatrix} -1 & 1 & 0 \\ 0 & -1 & 1 \\ 1 & 0 & -1 \end{bmatrix} \cdot \begin{bmatrix} 0,32 \\ 0,32 \\ 0,32 \end{bmatrix} = \begin{bmatrix} 0.0 \\ 0,0 \\ 0.0 \end{bmatrix}$$

It is worth observing that in the case of identical input data for the system of reservoirs and identical parameters related to WTPs, there are no transfers among reservoirs. The end reservoir filling level (for $W = 10\,[s]$) should be determined from Eq. 2.37), taking into account the values obtained for the vector of controlled outflows from reservoirs and the vector of transfers among reservoirs.

$$\hat{x}(W) = \begin{bmatrix} 30 \\ 30 \\ 30 \end{bmatrix}$$

The quality coefficient value in the discussed time interval $W = 10\,[s]$ for the assumed input data is: $F_{\min} = 6.144$, while the percentage execution of the water demand function for a given WTP is, respectively:

$$\mathbf{Y}(W) = \begin{bmatrix} 75\,\% & 0 & 0 & 0 \\ 0 & 75\,\% & 0 & 0 \\ 0 & 0 & 75\,\% & 0 \\ 0 & 0 & 0 & 75\,\% \end{bmatrix}$$

A graphical illustration of the example is shown in the diagrams (Fig. 2.4) obtained as a result of the system operation simulation carried out using an original application constructed in the Matlab/Simulink environment (Fig. 2.3). Each attempt to change the optimal control values obtained from the system operation simulation leads to an increase in the value of the assumed quality coefficient (2.10), or to a violation of the conditions imposed on reservoir end states (2.26). This proves the correctness of the solution, and ensures that it is optimum with regard to the coefficient and further assumptions applied.

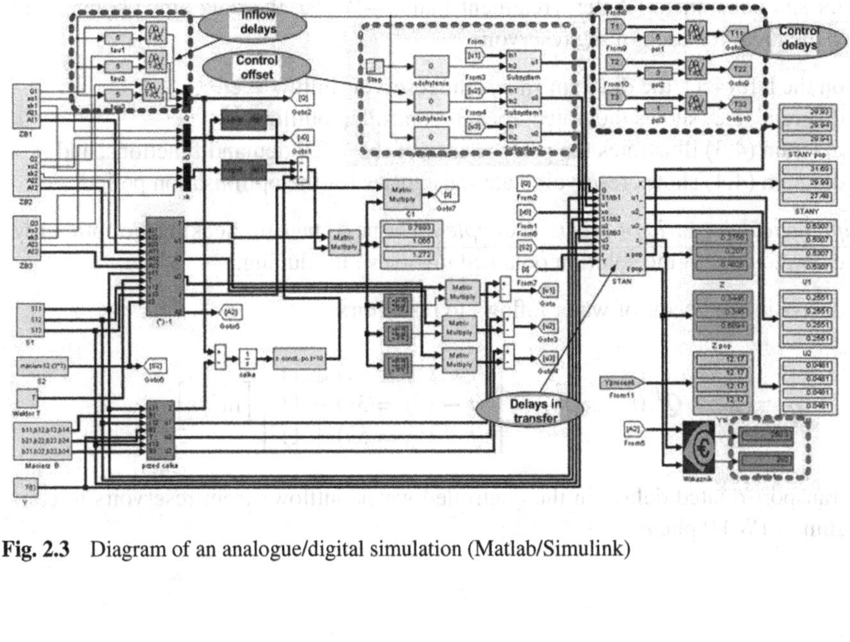

Fig. 2.3 Diagram of an analogue/digital simulation (Matlab/Simulink)

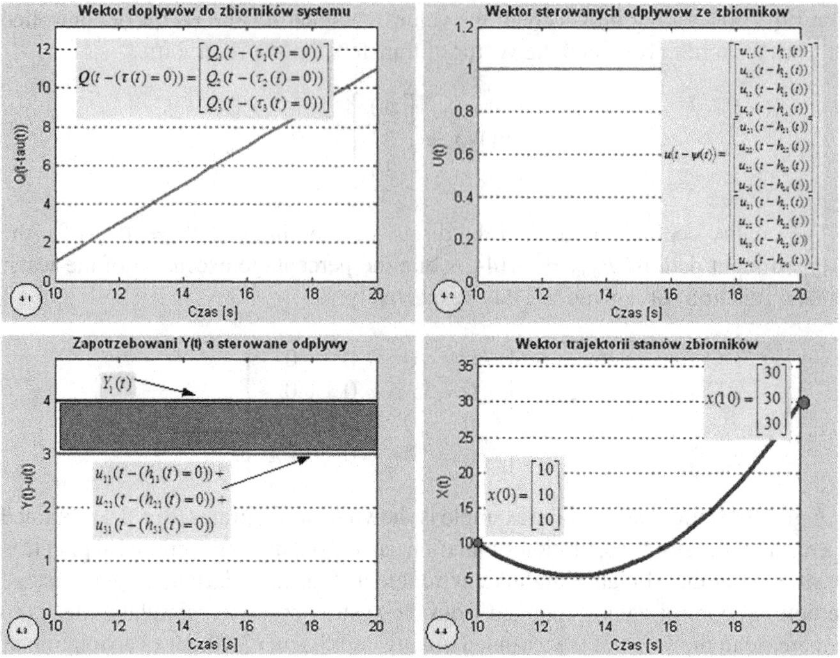

Fig. 2.4 Simulation results for a hypothetical scenario of events

Since the input data for the system operation simulation are identical for all reservoirs and consumers (Water Treatment Plants—WTP), the diagrams obtained are also identical for each of the reservoirs.

- on the left (4.1), the diagram shows the reservoir inflow vector,
- diagram (4.2) shows the trajectories of controlled outflows,
- diagram (4.3) illustrates the performance of the WTP demand functions, and
- diagram (4.4) shows reservoir state trajectories for the optimisation period

Hipothetical scenario of events. Example 2 The next stage in checking the optimality and correctness of the solution obtained involves introducing:

- delays in the vector of water inflows to reservoirs

$$\boldsymbol{Q}^P(t - \boldsymbol{\tau}(t)) = \begin{bmatrix} (t - (\tau_1 = 1s) + 1) \\ (t - (\tau_2 = 3s) + 1) \\ (t - (\tau_3 = 5s) + 1) \end{bmatrix} \left[\mathrm{m}^3/\mathrm{s}\right]$$

- transport-related delays in the controlled water outflows from reservoirs to consumers (WTP plants):

$$\hat{\boldsymbol{u}}(t-\boldsymbol{h}(t))_{(+)} = \begin{bmatrix} \begin{bmatrix} u_{11}(t-h_{11}(t)=5s) \\ u_{12}(t-h_{12}(t)=5s) \\ u_{13}(t-h_{13}(t)=5s) \\ u_{14}(t-h_{14}(t)=5s) \end{bmatrix} \\ \begin{bmatrix} u_{11}(t-h_{11}(t)=3s) \\ u_{12}(t-h_{12}(t)=3s) \\ u_{13}(t-h_{13}(t)=3s) \\ u_{14}(t-h_{14}(t)=3s) \end{bmatrix} \\ \begin{bmatrix} u_{11}(t-h_{11}(t)=1s) \\ u_{12}(t-h_{12}(t)=1s) \\ u_{13}(t-h_{13}(t)=1s) \\ u_{14}(t-h_{14}(t)=1s) \end{bmatrix} \end{bmatrix}$$

- other parameters remain unchanged

A graphical illustration of the example is shown in Fig. 2.5.

- Diagram (5.1) shows inflows to reservoirs taking into account delays during water flow through reservoirs.
- Diagram (5.2) contains the trajectories of controlled outflows as seen by the WTP water consumer. From this perspective, we see the impact of transport-related delays in water distribution from reservoirs to consumers.
- Diagram (5.3) clearly indicates that, for example, in different time intervals the demand function is carried out by 1, 2 or 3 reservoirs.
- Diagram (5.4) shows the significant trajectories of reservoir states, thus confirming that condition (26) is satisfied.
- Diagram (5.5) shows the trajectories of transfers among reservoirs.
- Diagram (5.6) demonstrates the directions of transfers among reservoirs (assumed and actual).

The quality coefficient value in the time interval discussed for assumed delays and input data is $F_{\min} = 84, 56$, while the percentage execution of water demand function for a WTP is, respectively:

$$\mathbf{Y}(W) = \begin{bmatrix} 12.08\,\% & 0 & 0 & 0 \\ 0 & 12.08\,\% & 0 & 0 \\ 0 & 0 & 12.08\,\% & 0 \\ 0 & 0 & 0 & 12.08\,\% \end{bmatrix}$$

Reduction of the quality coefficient value and percentage execution of the water demand function in WTP plants is dictated by the assumed transport-related delay in water delivery from the reservoirs to the WTPs. Water is distributed from reservoirs immediately as a result of receiving the optimising task solution, while, due to transport-related delays, the water is delivered to its consumer after a given time (with a delay) which, considering the quality coefficient, affects its value. This is so because, in different time intervals during the optimisation horizon, the WTP plants

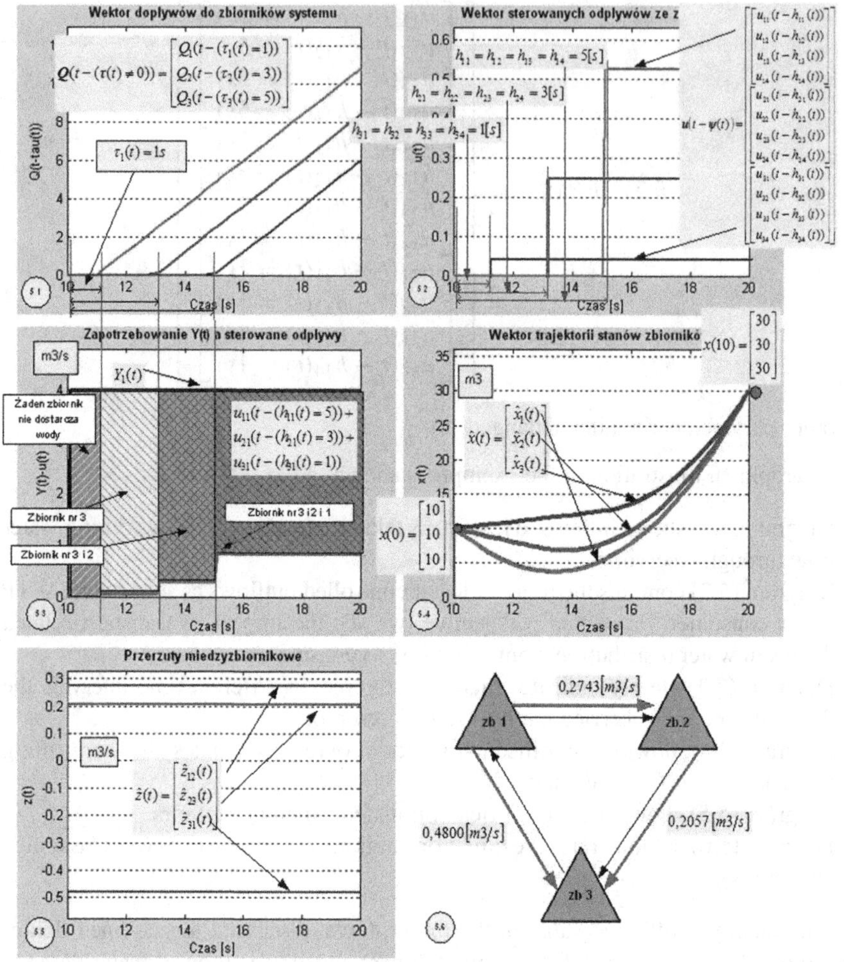

Fig. 2.5 Simulation results for a hypothetical scenario of events

either do not receive water at all or receive water from one, two or three reservoirs. In contrast, transport-related delays between reservoirs and water consumers (the WTP plants) have no impact on the end states in reservoirs and, as a result of this, condition (2.26) is always satisfied.

A completely different situation is observed when looking at the occurrence of delays during transfers among reservoirs. Delays in water transport among reservoirs affect the values of end states in reservoirs and, if they occur, condition (2.26) is not immediately satisfied. It is thus necessary to correct the values of water transfers among reservoirs in time, which does not affect the change in values of previously calculated controlled outflows from reservoirs to the WTP plants.

- examples of transport-related delays in water transfers among the system reservoirs, resulting from reservoir locations in the field;

$$\omega^{\mathrm{T}} = \left[\begin{bmatrix} \omega_{12} = 2s \\ \omega_{21} = 4s \end{bmatrix} \begin{bmatrix} \omega_{23} = 4s \\ \omega_{32} = 6s \end{bmatrix} \begin{bmatrix} \omega_{31} = 2.5s \\ \omega_{13} = 3s \end{bmatrix} \right]$$

other parameters remain unchanged.

A graphical illustration of the example is shown in Figs. 2.6 and 2.7. Diagrams 6.1, 6.2 and 6.3 are the same as Fig. 2.5. Diagram 6.4 shows the trajectories of transfers among reservoirs, in which delays in water transport among reservoirs have been taken into account. Diagram 6.5 presents the trajectories of reservoir states showing

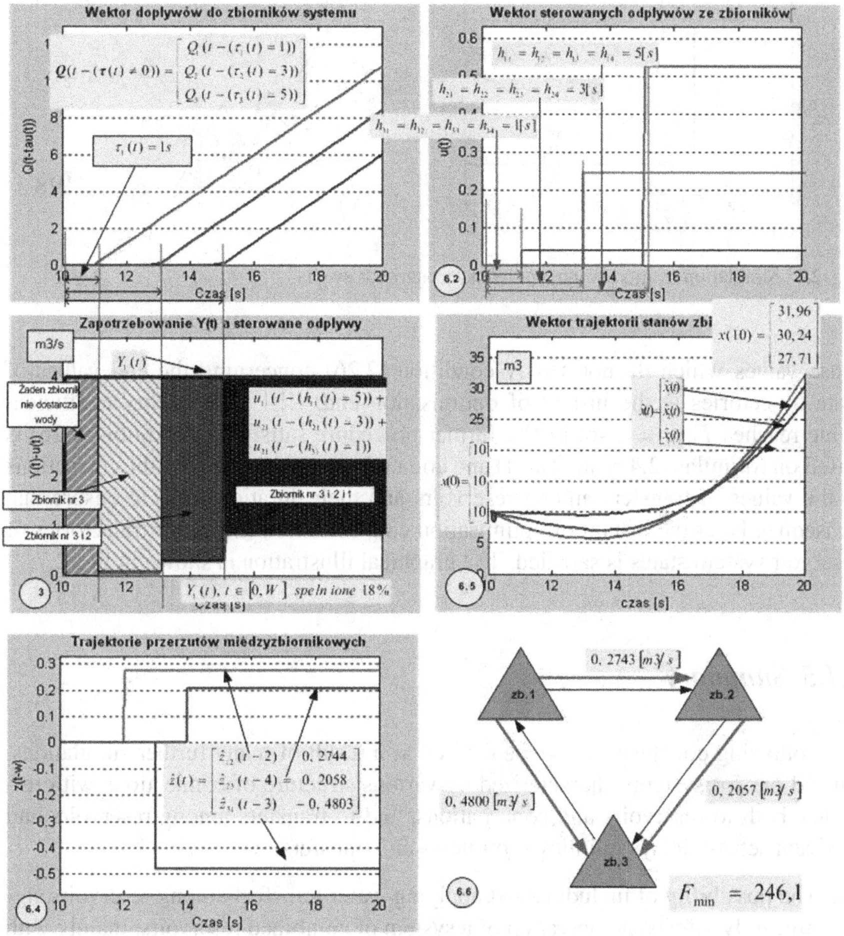

Fig. 2.6 Simulation results for a hypothetical scenario of events

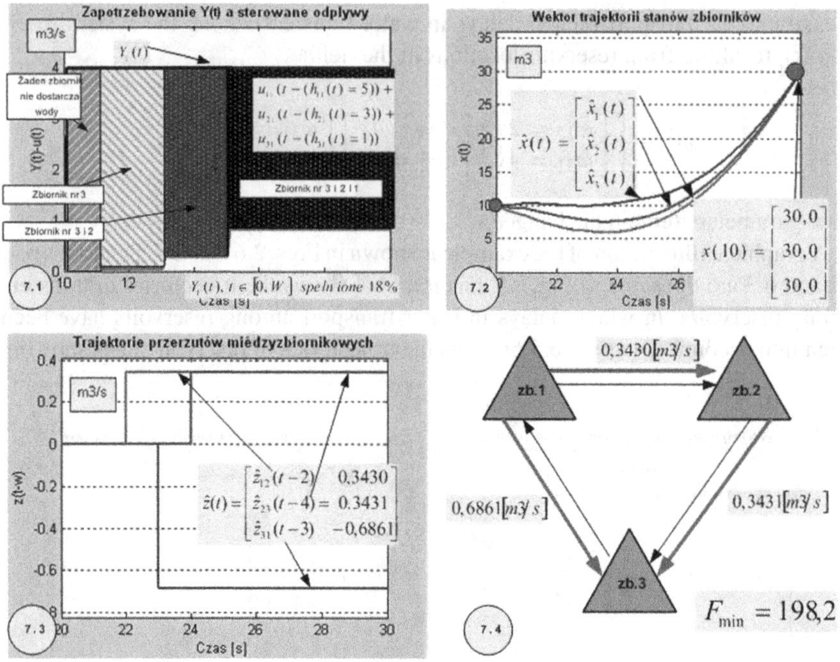

Fig. 2.7 Simulation results for a hypothetical scenario of events

those values which do not satisfy condition (2.26) concerning the end values of state trajectories at the instant of optimisation end. !!!!!. The quality coefficient value reaches $F_{\text{min}} = 246, 1$. The further procedure satisfying condition (2.26) is based on formulas (2.43) and (2.44) and comes down to determining the corrections to the values of transfers among reservoirs and their duration so as to ensure that consequently, at the instant of optimisation end, condition (2.26) for trajectories of reservoir system states is satisfied. The graphical illustration is shown in Fig. 2.7.

2.1.5 Summary

The following conclusions may be derived as a result of many further simulations, carried out for systems characterised by various structure of connections with reference both to reservoirs and conurbations, and to transfers among reservoirs and different sets of delays in inflows, outflows and transfers:

1. The possibility of including and applying water transfers among reservoirs significantly affects the operation of a system of combined reservoirs, mainly with respect to leaving the reservoir end states at the required levels (condition (2.26)), and that these states will be reached with a minimum value for the quality coeffi-

cient (2.10). The possibility to take into account delays related to water transfers among reservoirs considerably raises the substantive aspect of the solution.

2. Cooperation of a system of reservoirs in a configuration without transfers among reservoirs comes down to the operation of reservoirs which have one shared purpose—satisfy the water demands imposed on the system. None of the reservoirs *sees* other reservoirs in the system while satisfying the system's water demands assigned to that particular reservoir. In some cases this cooperation may lead to a situation in which, within a system of reservoirs working together, some of the reservoirs will remain with very low end states/levels with unfavourable, low predicted inflow and after optimisation time W. This unfavourable effect may be mitigated as a result of including transfers among reservoirs. According to the optimising task conditions, these transfers will be selected (for value and transfer direction) so as to ensure the required end states for the system reservoirs at a given vector of predicted inflows to the reservoir system.

Chapter 3
Transient Optimisation Horizon

3.1 Transient Optimisation Horizon

The search for an optimal optimisation duration may refer to cases of determining the following:

- free start time (optimal transient time for commencing optimisation, with a steady end time), (Free Start Time)
- free end time (optimal transient time for ending optimisation, with a known start time), (Free End Time).

In issues involving dispatch decision support for controlling outflows from reservoirs, the abovementioned cases of optimisation duration with combinations of boundary conditions for trajectories of states have broad applications. Due to the number of boundary condition variants, the moment of optimisation start or end is of fundamental importance. This article presents operation optimisation variants for a system of reservoirs with free optimisation time (Fig. 3.1):

- FST (Free Start Time)
- FET (Free End Time)

and the case of boundary conditions in the state trajectories in the following form:

- SLC (Steady Left Conditions in the state trajectories),
- SRC (Steady Right Conditions in the state trajectories).

Therefore, an integral part of the cases discussed will be to determine:

- the vector for optimal optimisation commencement times $\hat{\boldsymbol{t}}_{\boldsymbol{0}}^*$, for steady optimisation ending time W, or
- the vector for optimal optimisation commencement times $\hat{\boldsymbol{W}}^*$, for steady optimisation ending time t_0.

Because both start time and end time may be free, there are two characteristic cases in this regard, shown in Figs. 3.1 and 3.3. For the vector of predicted inflows to

W. Z. Chmielowski, *Management of Complex Multi-reservoir Water Distribution Systems Using Advanced Control Theoretic Tools and Techniques*, SpringerBriefs in Computational Intelligence, DOI: 10.1007/978-3-319-00239-2_3,

the reservoir system, the physical meaning of the quality coefficient (3.1) combines three tasks:

- to ensure the vector of outflows from reservoirs $\hat{\boldsymbol{u}}(t-\boldsymbol{h}(t))_{(+)}, \forall t \in \left[\hat{\boldsymbol{t}_0}^*, W\right]$ or $\left[t_0, \hat{\boldsymbol{W}}^*\right]$, which will minimally differ from the vector whose elements are partial water demands to be satisfied by the individual system $\mathbf{B}(t)\cdot\mathbf{Y}(t)\cdot\mathbf{S}\cdot\boldsymbol{1}, \forall t \in \left[\hat{\boldsymbol{t}_0}^*, W\right]$, or $\left[t_0, \hat{\boldsymbol{W}}^*\right]$
- to achieve the targets specified above at minimum transfer costs among reservoirs.
- to obtain (at the end of the optimisation horizon W, or $\hat{\boldsymbol{W}}^*$) the vector for reservoir filling levels satisfying the required values (correct boundary conditions in reservoir state trajectories).

3.2 Free Optimisation Time, FT

We have two cases here. The following may be searched for:

- optimal time for optimisation process commencement (free start time FST, and steady optimisation end time SET),
- optimal process ending time (steady start time SST, and free end time FET).

3.2.1 Quality Coefficient

For the reservoir system shown in Fig. 3.1, the quality coefficient for the task discussed assumes the following form:

$$F = 0,5\int_{\hat{\boldsymbol{t}_0}^*, t_0}^{W,\hat{\boldsymbol{W}}^*}\left\{\begin{array}{l}[\mathbf{B}(t)\cdot\mathbf{Y}(t)\cdot\mathbf{S}\cdot\boldsymbol{1}-\boldsymbol{u}(t-\boldsymbol{h}(t))]^{\mathrm{T}}_{(+)}\\ \cdot\mathbf{A}_1\cdot\\ [\mathbf{B}(t)\cdot\mathbf{Y}(t)\cdot\mathbf{S}\cdot\boldsymbol{1}-\boldsymbol{u}(t-\boldsymbol{h}(t))]_{(+)}\\ \cdot\mathbf{z}(t-\omega)^{\mathrm{T}}\cdot\mathbf{A}_2\cdot\mathbf{z}(t-\omega)\end{array}\right\}dt \tag{3.1}$$

where:

1. $\hat{\boldsymbol{t}_0}^*$ vector of free optimisation start times at W steady end time,
2. $\hat{\boldsymbol{W}}^*$ vector of free optimisation end times at t_0 steady start time,

Other symbols in coefficient (3.1) are the same as in (2.10).

3.2.2 The System State Equation

$$f:\dot{\boldsymbol{x}}(t)=\boldsymbol{Q}^P(t-\boldsymbol{\tau}(t))-\mathbf{S_1}\boldsymbol{u}(t-\boldsymbol{h}(t))+\mathbf{S_2}\boldsymbol{z}(t-\omega) \tag{3.2}$$

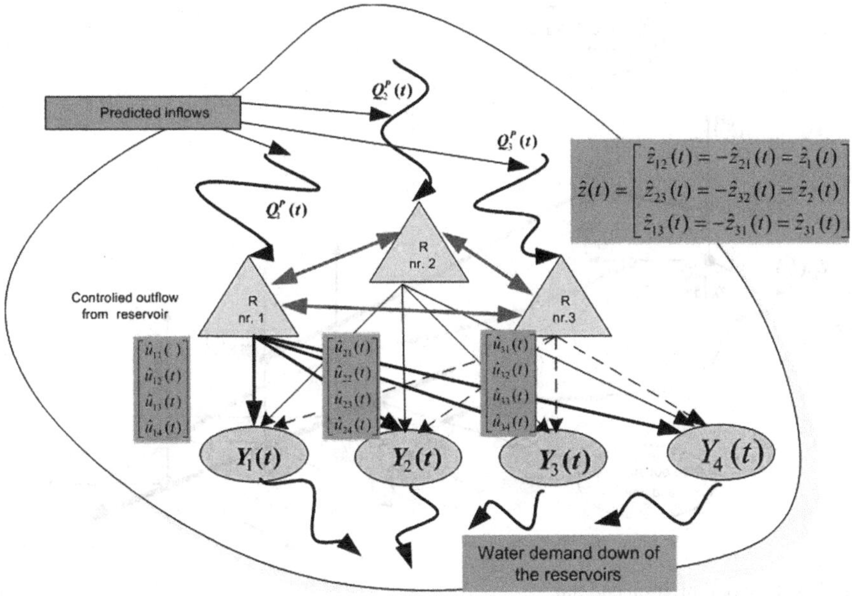

Fig. 3.1 The complex water-management system

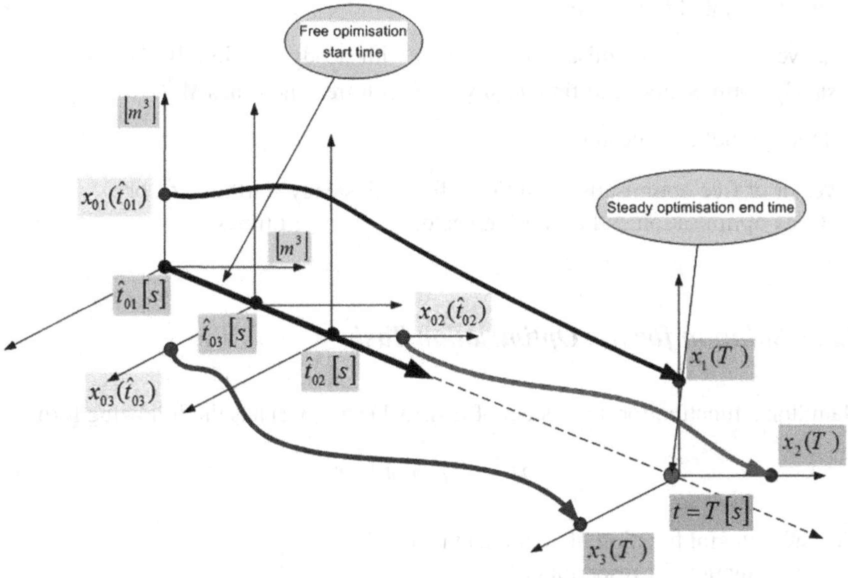

Fig. 3.2 Free optimisation start time

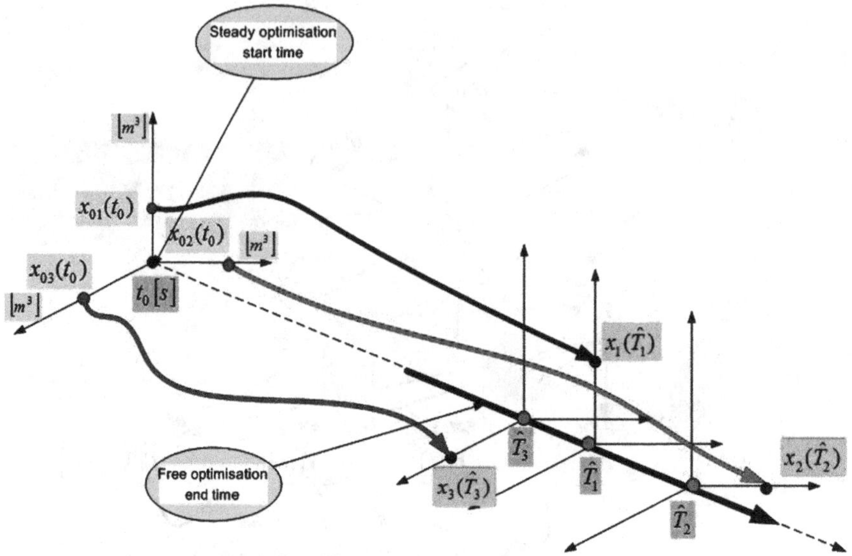

Fig. 3.3 Free optimisation end time

$$x(\#) = x_0 \qquad x(\&) = x_W \tag{3.3}$$

where the symbol # indicates:

- $\hat{\boldsymbol{t}}_{\boldsymbol{0}}^{*}$ vector of free optimisation start times with steady end time W, or
- steady optimisation start time t_0 at vector with free end times $\hat{\boldsymbol{W}}^{*}$

and the symbol & indicates:

- vector of free optimisation end times $\hat{\boldsymbol{W}}^{*}$ with steady start time t_0, or
- steady optimisation end time W at vector $\hat{\boldsymbol{t}}_{\boldsymbol{0}}^{*}$ free start times.

3.2.3 Solution for the Optimisation Task

Hamilton's function for the system of Eqs. (3.1) and (3.2) has the following form:

$$H = -f_0 + \boldsymbol{\psi}^{\mathrm{T}} \cdot \boldsymbol{f} \tag{3.4}$$

f_0—sub-integral function of coefficient (3.1)
$\boldsymbol{f}$—state equation for reservoirs (3.2)
$\boldsymbol{\psi}$—conjugate variable, vector [3*1]

$$H = -0,5 \cdot \left\{ \begin{array}{l} [\mathbf{B}(t) \cdot \mathbf{Y}(t) \cdot \mathbf{S} \cdot \mathit{1} - \boldsymbol{u}(t - \boldsymbol{h}(t))]^{\mathrm{T}}_{(+)} \\ \cdot \mathbf{A}_1 \cdot \\ [\mathbf{B}(t) \cdot \mathbf{Y}(t) \cdot \mathbf{S} \cdot \mathit{1} - \boldsymbol{u}(t - \boldsymbol{h}(t))]_{(+)} \\ \cdot \mathbf{z}(t - \omega)^{\mathrm{T}} \cdot \mathbf{A}_2 \cdot \mathbf{z}(t - \omega) \end{array} \right\}$$

$$+ \boldsymbol{\psi}(t)^{\mathrm{T}} \cdot \left[\boldsymbol{Q}^P(t - \boldsymbol{\tau}(t)) - \mathbf{S}_1 \boldsymbol{u}(t - \boldsymbol{h}(t)) + \mathbf{S}_2 \boldsymbol{z}(t - \omega) \right] \quad (3.5)$$

The system of equations for Hamilton's function in form (3.5) is shown below:

1.

$$\left[(\nabla_u H)_{\hat{u},\hat{x},\hat{\psi}} \right]^{\mathrm{T}} = \mathit{0} \Rightarrow \hat{\boldsymbol{u}}(t - \boldsymbol{h}(t))_{(+)}$$
$$= \mathbf{B}(t) \cdot \mathbf{Y}(t) \cdot \mathbf{S} \cdot \mathit{1} - \mathbf{A}_1^{-1} \cdot \mathbf{S}_1{}^{\mathrm{T}} \cdot \hat{\boldsymbol{\psi}}(t) \quad (3.6)$$

2.

$$\left[(\nabla_z H)_{\hat{u},\hat{z},\hat{x},\hat{\psi}} \right]^{\mathrm{T}} = \mathit{0} \Rightarrow \hat{\mathbf{z}}(t - \omega) = \mathbf{A}_2^{-1} \cdot \mathbf{S}_2^{\mathrm{T}} \cdot \hat{\boldsymbol{\psi}}(t) \quad (3.7)$$

3.

$$\left[(\nabla_\psi H)_{\hat{u},\hat{x}} \right]^{\mathrm{T}} = \dot{\boldsymbol{x}}(t)$$
$$\Rightarrow \boldsymbol{Q}^P(t - \boldsymbol{\tau}(t)) - \mathbf{S}_1 \boldsymbol{u}(t - \boldsymbol{h}(t)) + \mathbf{S}_2 \boldsymbol{z}(t - \omega) \quad (3.8)$$

4.

$$\left[-(\nabla_x H)_{\hat{u},\hat{x},\hat{\psi}} \right]^{\mathrm{T}} = \dot{\hat{\boldsymbol{\psi}}}(t) \Rightarrow \dot{\hat{\boldsymbol{\psi}}}(t) = \mathit{0}_{(3*1)} \quad (3.9)$$

using Eq. (3.6) through (3.9), we determine the vector of constants $\boldsymbol{C}_1$.

3.2.4 For the Vector of Free Optimisation Start Times

$$\boldsymbol{C}_1 = \left[\left(\mathbf{S}_1 \cdot \mathbf{A}_1^{-1} \cdot \mathbf{S}_1^{\mathrm{T}} + \mathbf{S}_2 \cdot \mathbf{A}_2^{-1} \cdot \mathbf{S}_2^{\mathrm{T}} \right) \cdot \left(W - \hat{\boldsymbol{t}_0}^* \right) \right]^{-1}$$
$$\cdot \left\{ \begin{array}{l} \boldsymbol{x}(W) - \boldsymbol{x}(\hat{\boldsymbol{t}_0}^*) - \\ + \int_{\hat{\boldsymbol{t}_0}^*}^{W} \left[\boldsymbol{Q}^P(t - \boldsymbol{h}(t)) - \mathbf{S}_1 \cdot \mathbf{B}(t) \cdot \mathbf{Y}(t) \cdot \mathbf{S} \cdot \mathit{1} \right] dt \end{array} \right\} \quad (3.10)$$

The vector of constants $\boldsymbol{C}_1$ is dependent on the vector of free optimisation start times $\hat{\boldsymbol{t}_0}^*$ and to compute it we need another equation, which may be derived from the following relation:

$$H(\hat{t_0}^*) - \left(\frac{\partial K \left[x(\hat{t_0}^*), \hat{t_0}^* \right]}{\partial \hat{t_0}^*} \right) = 0 \tag{3.11}$$

In this case, coefficient (3.1) does not contain the functions of initial conditions $K\left[x(\hat{t_0}^*), \hat{t_0}^*\right]$, therefore the Hamiltonian function (3.4) $H(\hat{t_0}^*) = 0$. Further, we convert the Hamiltonian function, aiming at determining the $\boldsymbol{\psi}(\hat{t_0}^*)$

$$0 = -0,5 \cdot \left\{ \begin{array}{l} [\mathbf{B}(t) \cdot \mathbf{Y}(t) \cdot \mathbf{S} \cdot \mathit{1} - \boldsymbol{u}(t - \boldsymbol{h}(t))]^{\mathrm{T}}_{(+)} \\ \cdot \mathbf{A}_1 \cdot \\ \cdot \mathbf{z}(t-\omega)^{\mathrm{T}} \cdot \mathbf{A}_2 \cdot \mathbf{z}(t-\omega) \end{array} \right\}$$
$$+ \boldsymbol{\psi}(t)^{\mathrm{T}} \cdot \left[\boldsymbol{Q}^P(t - \boldsymbol{\tau}(t)) - \mathbf{S}_1 \boldsymbol{u}(t - \boldsymbol{h}(t)) + \mathbf{S}_2 \boldsymbol{z}(t-\omega) \right] \tag{3.12}$$

We substitute term (3.6) for $\hat{\boldsymbol{u}}(t - \boldsymbol{h}(t))_{(+)}$ and for $t = \hat{t_0}^*$ we rearrange it to the form (3.13):

$$-0,5 \cdot \left\{ \begin{array}{l} \left[\mathbf{A}_1^{-1} \cdot \mathbf{S}_1^{\mathrm{T}} \cdot \hat{\boldsymbol{\psi}}(\hat{t_0}^*) \right]^{\mathrm{T}}_{(+)} \\ \left[\mathbf{A}_1^{-1} \cdot \mathbf{S}_1^{\mathrm{T}} \cdot \hat{\boldsymbol{\psi}}(\hat{t_0}^*) \right]_{(+)} + \\ \left[\mathbf{A}_2^{-1} \cdot \mathbf{S}_2^{\mathrm{T}} \cdot \hat{\boldsymbol{\psi}}(\hat{t_0}^*) \right]^{\mathrm{T}} \\ \cdot \mathbf{A}_2 \cdot \left[\mathbf{A}_2^{-1} \cdot \mathbf{S}_2^{\mathrm{T}} \cdot \hat{\psi}(\hat{t_0}^*) \right] \end{array} \right\}$$
$$+ \hat{\boldsymbol{\psi}}(\hat{t_0}^*)^{\mathrm{T}} \cdot \left[\begin{array}{l} \boldsymbol{Q}^P((\hat{t_0}^*) - \boldsymbol{\tau}(t_0^*)) - \\ + \mathbf{S}_1 \cdot \mathbf{B}(\hat{t_0}^*) \cdot \mathbf{Y}(\hat{t_0}^*) \cdot \mathbf{S} \cdot \mathit{1} \\ + \mathbf{S}_1 \cdot \mathbf{A}_1^{-1} \cdot \mathbf{S}_1^{\mathrm{T}} \cdot \hat{\boldsymbol{\psi}}(\hat{t_0}^*) \\ + \mathbf{S}_2 \cdot \mathbf{A}_2^{-1} \cdot \mathbf{S}_2^{\mathrm{T}} \cdot \hat{\boldsymbol{\psi}}(\hat{t_0}^*) \end{array} \right] = 0 \tag{3.13}$$

After further multiplications, we receive (3.14):

$$\begin{array}{l} -0,5 \cdot \hat{\boldsymbol{\psi}}(\hat{t_0}^*))^{\mathrm{T}} \cdot \mathbf{S}_1 \cdot \mathbf{A}_1^{-1} \cdot \mathbf{S}_1^{\mathrm{T}} \cdot \hat{\boldsymbol{\psi}}(\hat{t_0}^*) + \\ -0,5 \cdot \hat{\boldsymbol{\psi}}(\hat{t_0}^*)^{\mathrm{T}} \cdot \mathbf{S}_2 \cdot \mathbf{A}_2^{-1} \cdot \mathbf{S}_2^{\mathrm{T}} \cdot \hat{\boldsymbol{\psi}}(\hat{t_0}^*) \\ + \hat{\boldsymbol{\psi}}(\hat{t_0}^*)^{\mathrm{T}} \cdot \boldsymbol{Q}^P((\hat{t_0}^*) - \boldsymbol{\tau}(\hat{t_0}^*)) - \\ + \hat{\boldsymbol{\psi}}(\hat{t_0}^*)^{\mathrm{T}} \cdot \mathbf{S}_1 \cdot \mathbf{B}(\hat{t_0}^*) \cdot \mathbf{Y}(\hat{t_0}^*) \cdot \mathbf{S} \cdot \mathit{1} \\ + \hat{\boldsymbol{\psi}}(\hat{t_0}^*)^{\mathrm{T}} \cdot \mathbf{S}_1 \cdot \mathbf{A}_1^{-1} \cdot \mathbf{S}_1^{\mathrm{T}} \cdot \hat{\boldsymbol{\psi}}(\hat{t_0}^*) \\ + \hat{\boldsymbol{\psi}}(\hat{t_0}^*)^{\mathrm{T}} \cdot \mathbf{S}_2 \cdot \mathbf{A}_2^{-1} \cdot \mathbf{S}_2^{\mathrm{T}} \cdot \hat{\boldsymbol{\psi}}(\hat{t_0}^*) = 0 \end{array} \tag{3.14}$$

We divide equation (3.14) by $\hat{\boldsymbol{\psi}}(\hat{t_0}^*)^T$ (left):

$$\begin{aligned}
&-0,5\cdot \mathbf{S}_1\cdot \mathbf{A}_1^{-1}\cdot \mathbf{S}_1^T\cdot \hat{\boldsymbol{\psi}}(\hat{t_0}^*)+\\
&-0,5\cdot \mathbf{S}_2\cdot \mathbf{A}_2^{-1}\cdot \mathbf{S}_2^T\cdot \hat{\boldsymbol{\psi}}(\hat{t_0}^*)+\\
&+\boldsymbol{Q}^P((\hat{t_0}^*)-\boldsymbol{\tau}(\hat{t_0}^*))-\\
&+\mathbf{S}_1\cdot \mathbf{B}(\hat{t_0}^*)\cdot \mathbf{Y}(\hat{t_0}^*)\cdot \mathbf{S}\cdot \boldsymbol{1}+\\
&+\mathbf{S_1}\cdot \mathbf{A}_1^{-1}\cdot \mathbf{S}_1^{\mathrm{T}}\cdot \hat{\boldsymbol{\psi}}(\hat{t_0}^*)+\\
&+\mathbf{S_2}\cdot \mathbf{A}_2^{-1}\cdot \mathbf{S}_2^{\mathrm{T}}\cdot \hat{\boldsymbol{\psi}}(\hat{t_0}^*)=0
\end{aligned} \tag{3.15}$$

After regrouping Eq. (3.15) we receive:

$$\begin{aligned}
&0,5\cdot\left[\mathbf{S}_1\cdot \mathbf{A}_1^{-1}\cdot \mathbf{S}_1^{\mathrm{T}}+\mathbf{S_2}\cdot \mathbf{A}_2^{-1}\cdot \mathbf{S}_2^{\mathrm{T}}\right]\cdot \hat{\boldsymbol{\psi}}(\hat{t_0}^*)\\
&=\mathbf{S}_1\cdot \mathbf{B}(\hat{t_0}^*)\cdot \mathbf{Y}(\hat{t_0}^*)\cdot \mathbf{S}\cdot \boldsymbol{1}-\boldsymbol{Q}^P((\hat{t_0}^*)-\boldsymbol{\tau}(\hat{t_0}^*))
\end{aligned} \tag{3.16}$$

and finally, $\hat{\boldsymbol{\psi}}(\hat{t_0}^*)$ is:

$$\begin{aligned}
\hat{\boldsymbol{\psi}}(\hat{t_0}^*)&=0,5\cdot\left[\mathbf{S}_1\cdot \mathbf{A}_1^{-1}\cdot \mathbf{S}_1^{\mathrm{T}}+\mathbf{S_2}\cdot \mathbf{A}_2^{-1}\cdot \mathbf{S}_2^{\mathrm{T}}\right]^{-1}\\
&\cdot\left[\mathbf{S}_1\cdot \mathbf{B}(\hat{t_0}^*)\cdot \mathbf{Y}(\hat{t_0}^*)\cdot \mathbf{S}\cdot \boldsymbol{1}-\boldsymbol{Q}^P((\hat{t_0}^*)-\boldsymbol{\tau}(\hat{t_0}^*))\right]
\end{aligned} \tag{3.17}$$

Further problem solving proceeds as follows:

- By comparing Eqs. (3.10) and (3.16), it is possible to determine the vector for optimal process starting times.
- Knowing optimisation process start time and using formula (3.3), we can determine the vector for the initial filling levels for the reservoirs $t = \hat{t}_i^*,\ i = 1, \ldots, 3 \Rightarrow x_{0,i} = x_{0,i}(\hat{t}_i^*)$.
- Using (3.17), we determine the vector of conjugate variables $\hat{\boldsymbol{\psi}}(\hat{t_0}^*) = \mathbf{C}_1$.
- Knowing the vector of conjugate variables, according to (3.6), we can determine the optimal control taking into account transport-related delays between reservoirs and water consumers (WTP plants).
- Knowing the vector of conjugate variables, according to (3.7), we can determine the vector of transfers among reservoirs skipping transport-related delays.
- As a result of taking into account the delays related to transfers among reservoirs, we upset the condition for the final filling levels in reservoirs, and then $\boldsymbol{x}(W) \neq \boldsymbol{x}(W)^U$, condition (2.26). Modifying the value of the vector of transfers among reservoirs comes down to procedure formula (2.43 and 2.44), after application of which $\boldsymbol{x}(W) = \boldsymbol{x}(W)^U$.
- We receive a minimum value for the quality coefficient when substituting $\hat{\boldsymbol{u}}(t - \boldsymbol{h}(t))_{(+)}$ and $\hat{\mathbf{z}}(t - \boldsymbol{\omega})$ to (3.1).

3.2.5 Computer Simulation

In the example, we assume the following input data with reference to reservoirs and water consumers:

- initial/start states in reservoirs $\left[m^3\right]$ change in function of optimisation start time:

$$\boldsymbol{x}(\boldsymbol{t_0})^{\mathrm{T}} = \left[\, 10 - t_{0_{res1}} \;\; 20 - 0.5t_{0_{res2}} \;\; 30 + 0.2t_{0_{res3}} \,\right]$$

- the following final/end states in reservoirs $\left[m^3\right]$ have been assumed:

$$\boldsymbol{x}(W)^{\mathrm{T}} = \left[30\ 40\ 50\right]$$

- the following inflows to reservoirs $\left[m^3\right]$ have been assumed:

$$\boldsymbol{Q}(t)^{\mathrm{T}} = \left[(t+1)\ (0.5t+2)\ (0.4t+3)\right]$$

- water demand $\left[m^3/s\right]$ below reservoirs:

$$\mathbf{Y}(t) = \begin{bmatrix} [\otimes] & * & * \\ * & [\otimes] & * \\ * & * & [\otimes] \end{bmatrix}$$

$$[\otimes] = \begin{bmatrix} 2+0.1t & 0 & 0 & 0 \\ 0 & 3\cdot 1(t) & 0 & 0 \\ 0 & 0 & 3.5+0.2t & 0 \\ 0 & 0 & 0 & 4-0.1t \end{bmatrix}$$

- the functions for reservoir involvement in carrying out the demand functions:

$$\mathbf{B}(t) = \begin{bmatrix} [\otimes_1] & * & * \\ * & [\otimes_2] & * \\ * & * & [\otimes_3] \end{bmatrix},$$

$$[\otimes_1] = \begin{bmatrix} 0.2 & 0 & 0 & 0 \\ 0 & 0.4 & 0 & 0 \\ 0 & 0 & 0.4 & 0 \\ 0 & 0 & 0 & 0.3 \end{bmatrix} \cdot (t)$$

$$[\otimes_2] = \begin{bmatrix} 0.4 & 0 & 0 & 0 \\ 0 & 0.3 & 0 & 0 \\ 0 & 0 & 0.3 & 0 \\ 0 & 0 & 0 & 0.4 \end{bmatrix} \cdot (t)$$

$$[\otimes_3] = \begin{bmatrix} 0.4 & 0 & 0 & 0 \\ 0 & 0.3 & 0 & 0 \\ 0 & 0 & 0.3 & 0 \\ 0 & 0 & 0 & 0.3 \end{bmatrix} \cdot (t)$$

- matrixes of weight coefficients:

$$\mathbf{A}_1 = \begin{bmatrix} [\oplus] & * & * \\ * & [\oplus] & * \\ * & * & [\oplus] \end{bmatrix}$$

$$[\oplus] = \begin{bmatrix} 1\,0\,0\,0 \\ 0\,1\,0\,0 \\ 0\,0\,1\,0 \\ 0\,0\,0\,1 \end{bmatrix} \quad \mathbf{A}_2 = \begin{bmatrix} 1\,0\,0 \\ 0\,1\,0 \\ 0\,0\,1 \end{bmatrix}$$

- structural matrix

$$\mathbf{S}_1 = \begin{bmatrix} [\times] & \circ & \circ \\ \circ & [\times] & \circ \\ \circ & \circ & [\times] \end{bmatrix},$$

$$[\times] = \begin{bmatrix} 1\,1\,1\,1 \end{bmatrix}$$

$$[\circ] = \begin{bmatrix} 0\,0\,0\,0 \end{bmatrix}$$

- structural matrix

$$\mathbf{S}_2 = \begin{bmatrix} -1 & 0 & 1 \\ 1 & -1 & 0 \\ 0 & 1 & -1 \end{bmatrix}$$

- end time $W = 10\ [s]$

Graphical illustration of the example prepared using an original simulating application (Matlab/Simulink environment—Fig. 3.4) is shown in Fig. 3.5.

The left column in Fig. 3.5 presents:

1. inflows to reservoir (first diagram),
2. controlled outflows from reservoirs to the benefit of successive conurbations (diagrams 2, 3, 4),
3. controlled transfers among reservoirs (diagram 5). Black indicates the transfer directed (supply reservoir, receiving reservoir) according to the direction assumed in the system state equation. Red indicates those transfers whose direction is opposite to that assumed in the state equation.

The right columns show the following comparisons:

1. Sums of outflows from reservoirs with reference to the successive demand functions of WTP stations. Due to the varying input data for the reservoirs, operation

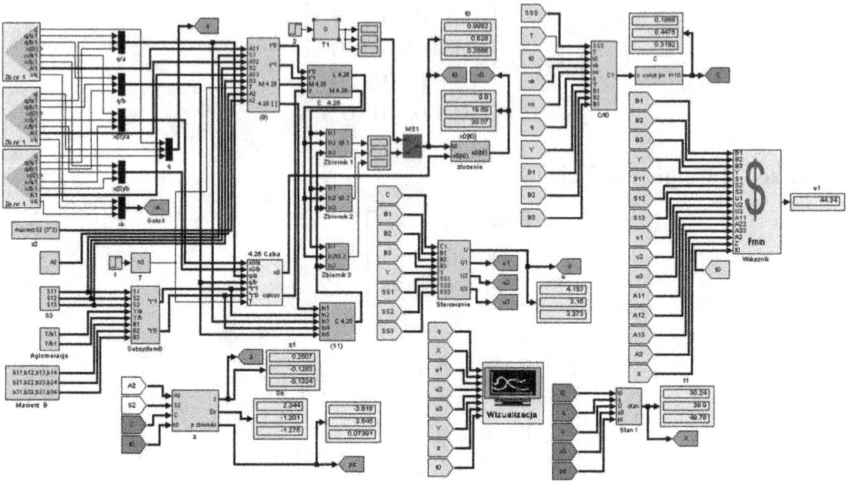

Fig. 3.4 Analogue/digital model

times for individual reservoirs are not the same. The shaded fields show the differences among demand functions Y_i, $i = 1, \ldots, 4$ for WTP stations, and that they are satisfied by controlled outflows from reservoirs (diagrams 6, 7, 8, 9),

2. Diagram (10) represents the reservoir state trajectories with marked optimal times for adding individual reservoirs to the whole system operation. Operation times dete-rmine the initial optimum state value for individual reservoirs in the light of (3.1).

The conclusions arrived at after analysis of the aforementioned issue may be formulated as follows:

- as regards the issue presented (free optimisation start time), the times for adding successive reservoirs to the whole system operation are generally different and conditioned by the input data for individual reservoirs (forecast of water inflow to the reservoir, reservoir involvement in executing the demand function vector, delays in water inflows to the system, and delays in water distribution from the system). Weight coefficients are also of importance (matrixes $\mathbf{A_1}$, $\mathbf{A_2}$).
- If the initial filling level states are functions of their respective operation times, then determined optimal times for adding individual reservoirs to the whole system operation directly affect the initial state values in the reservoirs.
- Among other things, the role of transfers among reservoirs comes down to minimising the differences between the assumed and actual end states of the reservoirs.
- Each change in the results obtained (controlled outflows, operation times) results in an increase in the value of coefficient (3.1).

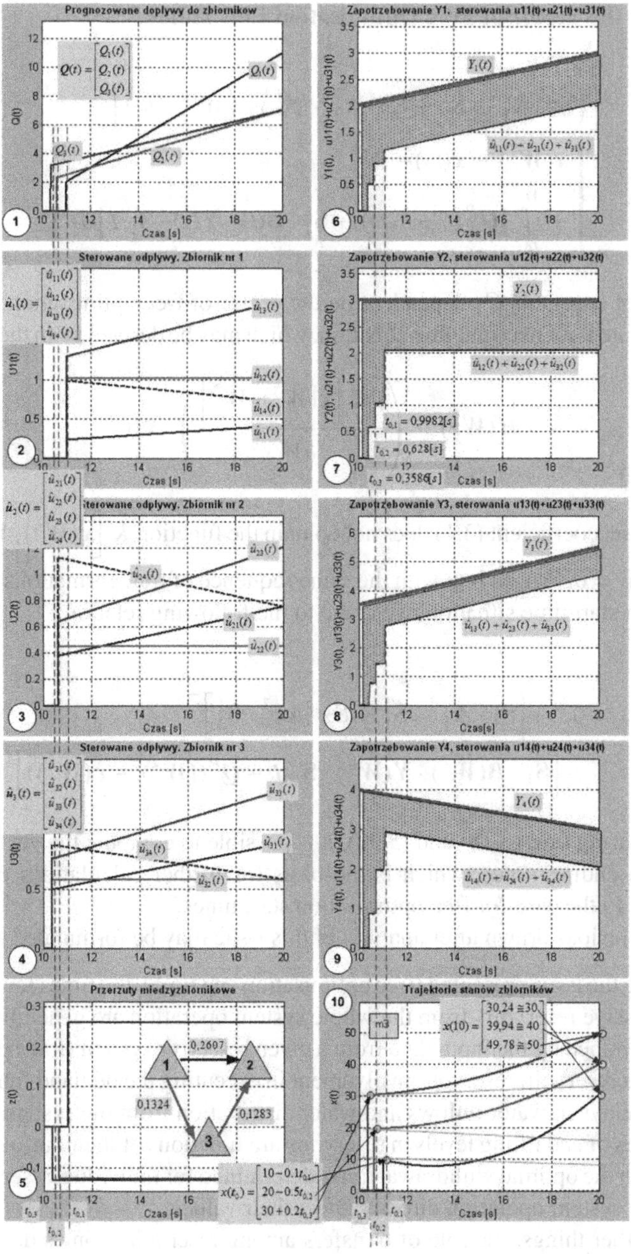

Fig. 3.5 Simulation results for the assumed input data

3.2.5.1 For the Vector of Free Optimisation End Times

$$C_1 = \left[\left(\mathbf{S}_1 \cdot \mathbf{A}_1^{-1} \cdot \mathbf{S}_1^{\mathrm{T}} + \mathbf{S}_2 \cdot \mathbf{A}_2^{-1} \cdot \mathbf{S}_2^{\mathrm{T}}\right) \cdot \left(\hat{\boldsymbol{W}}^* - t_0\right)\right]^{-1} \cdot \left\{ \begin{array}{l} \boldsymbol{x}(\hat{\boldsymbol{W}}^*) - \boldsymbol{x}(t_0) - \\ + \int\limits_{t_0}^{\hat{\boldsymbol{W}}^*} \left[\boldsymbol{Q}^P(t - \boldsymbol{h}(t)) - \mathbf{S}_1 \cdot \mathbf{B}(t) \cdot \mathbf{Y}(t) \cdot \mathbf{S} \cdot \boldsymbol{1}\right] dt \end{array} \right\} \tag{3.18}$$

The vector of constants $\boldsymbol{C}_1$ dependent on the vector of free optimisation end times, and to compute it we need another equation, which may be derived from the following relation:

$$H(\hat{\boldsymbol{W}}^*) - \left(\frac{\partial K\left[\boldsymbol{x}(\hat{\boldsymbol{W}}^*),\, \hat{\boldsymbol{W}}^*\right]}{\partial \hat{\boldsymbol{W}}^*} \right) = 0 \tag{3.19}$$

In this case, coefficient (3.1) does not contain the function $K\left[\boldsymbol{x}(\hat{\boldsymbol{W}}^*),\, \hat{\boldsymbol{W}}^*\right]$ therefore $H(\hat{\boldsymbol{W}}^*) = 0$. Proceeding with the same sequence of transformations as the free optimisation start time situation, we come to the following relation:

$$\hat{\boldsymbol{\psi}}(\hat{\boldsymbol{W}}^*) = 0,5 \cdot \left[\mathbf{S}_1 \cdot \mathbf{A}_1^{-1} \cdot \mathbf{S}_1^{\mathrm{T}} + \mathbf{S}_2 \cdot \mathbf{A}_2^{-1} \cdot \mathbf{S}_2^{\mathrm{T}}\right]^{-1} \cdot \left[\mathbf{S}_1 \cdot \mathbf{B}(\hat{\boldsymbol{W}}^*) \cdot \mathbf{Y}(\hat{\boldsymbol{W}}^*) \cdot \mathbf{S} \cdot \boldsymbol{1} - \boldsymbol{Q}^P((\hat{\boldsymbol{W}}^*) - \boldsymbol{\tau}(\hat{\boldsymbol{W}}^*))\right] \tag{3.20}$$

By comparing Eqs. (3.18) and (3.20), it is possible to compute the vector of optimal process ending times, which is the basis for further calculations carried out analogously to the case for free optimisation start times.

The conclusions drawn after analysis of this issue may be formulated as follows:

- Regarding the issue presented (free optimisation end time), the times for withdrawing successive reservoirs from the whole system operation are generally different and conditioned by the input data from a given reservoir system (forecast of water inflow to a reservoir, reservoir involvement in executing the demand function vector, and delays in water inflows and water distribution from the system).
- If the states of end filling levels in reservoirs are functions of their out-of-operation times, then the optimal withdrawal times determined for individual reservoirs from the whole system operation directly affect their values.
- Among other things, the role of transfers among reservoirs comes down to minimising the differences between the reservoirs' assumed and actual end states.
- Each change in the results obtained (controlled outflows, out-of-operation times) results in an increased value for coefficient (3.1).

3.3 Summary

This part discusses the issues involved in controlling a system of storage/retention reservoirs which supply water to a certain group of consumers. The issues presented may be cases frequently occurring in practice. Steady start and end conditions in state trajectories constitute a very practical optimisation task which may form the basis for the functional control of complex reservoir systems. Additionally, transient optimisation start/end time and the introduction of delays related to water inflow to reservoirs and its distribution from the system significantly increase the selection of possible cases. The numerical examples presented in this article apply to a specific system structure (reservoirs, consumers, links). In contrast, the analytical formulas obtained as a result of solving specified tasks are general formulas. They are valid in a fully optional dimensionality with reference to the system elements and links among them, which allows the analysis of the work of freely configured water distribution systems. Original computer applications have been used to perform the computations and then to discuss the simulation results for operating a model system using hypothetical data forming the scenarios for events within the system being analysed. In consequence, it is possible to propose various control variants (dependent on the scenario) which have one characteristic in common—they treat a forecast in a deterministic way.

In each case, simulation results include:

1. vector of optimal optimisation start/end times,
2. vector of outflows from reservoirs for the whole defined optimisation horizon,
3. vector of transfers among reservoirs,
4. vector for final filling levels in system reservoirs which guarantees the water volume required in the system reservoirs at the end of discussed time horizon.

Analysis of an adequately large set of hypothetical situations allows effective determination of the system structure and variability range for its major parameters $\mathbf{B}(t), \mathbf{Y}(t), \mathbf{A}_1, \mathbf{A}_2, \mathbf{A}_3, \boldsymbol{x}(W)$. Obtained solutions: control vector $\hat{\boldsymbol{u}}(t - \boldsymbol{h}(t))_{(+)}$, vector of transfers among reservoirs $\hat{z}(t)$, and in consequence the value of quality coefficient F are the functions of these parameters.

Part II
Related Boundary Conditions in the Trajectories of States for Optimal Management of Complex Multi-Reservoir Water Distribution System

Chapter 4
Introduction

Left and/or right boundary conditions in the trajectories of states may be free, bounded or steady. In this article we will focus on related conditions in the trajectories of states, meaning that e.g. initial and/or final water levels in reservoirs have to satisfy binding equations, which are determined by the requirements set for the system of reservoirs. In order to illustrate this we will take simple equations of planes changing in time, in which points correspond to total water volume in all system reservoirs.

- Equation of plane for initial conditions $g_1(t_0^*)$, indicates optimal time to start reservoir system operation. Steady end time W

$$\left[d_1(t_0^*) \bullet d_m(t_0^*) \right] \cdot \begin{bmatrix} x_1(t_0^*) \\ \bullet \\ x_m(t_0^*) \end{bmatrix} - b_1(t_0^*) = 0 \tag{4.1}$$

- Equation of plane for final conditions $g_2(\widehat{W})$ in which $\widehat{W}$ indicates optimal time to end reservoir system operation. Steady start time t_0

$$\left[e_1(\widehat{W}) \bullet e_m(\widehat{W}) \right] \cdot \begin{bmatrix} x_1(\widehat{W}) \\ \bullet \\ x_m(\widehat{W}) \end{bmatrix} - b_2(\widehat{W}) = 0 \tag{4.2}$$

Optimisation time may be steady or free, with reference to both start and end optimisation time. As regards free optimisation start time, there are two further issues:

- Free Start Time (FST) t_0^*, at steady optimisation end time W.
- Free End Time (FET) $\hat{W}$, at steady optimisation start time t_0.

W. Z. Chmielowski, *Management of Complex Multi-reservoir Water Distribution Systems Using Advanced Control Theoretic Tools and Techniques*, SpringerBriefs in Computational Intelligence, DOI: 10.1007/978-3-319-00239-2_4,

Chapter 1
Introduction

Chapter 5
Free Optimisation Time, FT

5.1 Free Optimisation Time, FT

Figure 5.1 presents a model system m of combined reservoirs which supply water to n consumers who are independent of each other. The final conditions in the reservoir state trajectories are linked by Eqs. (4.1) and (4.2), which show that at the beginning of the optimisation horizon t_0^* the sum of water volumes in the system reservoirs is to be satisfied by Eq. (4.1), and after optimisation period $\hat{W}$ the total volume of reservoir states is to be satisfies by Eq. (4.2). For a three-dimensional space, conditions (4.1) and (4.2) are demonstrated in Fig. 5.2. For the vector of predicted inflows to reservoirs, the optimisation task formulated by index (5.11) is reduced to:

- determining the control vector (for outflows from reservoirs) in time interval $\left[t_0^*, W\right]$ (free start time),

$$\hat{\boldsymbol{u}}(t - \boldsymbol{h}(t))_{(+)}, \quad \forall t \in \left[t_0^*, W\right] \tag{5.1}$$

  or in time interval $\left[t_0, \hat{W}\right]$ free optimisation end time

$$\hat{\boldsymbol{u}}(t - \boldsymbol{h}(t))_{(+)}, \quad \forall t \in \left[t_0, \hat{W}\right] \tag{5.2}$$

  which will minimally differ from the vector composed of partial water demands per individual system reservoirs,

$$\mathbf{B}(t) \cdot \mathbf{Y}(t) \cdot \mathbf{S} \cdot \boldsymbol{1}, \quad \forall t \in \left[t_0^*, W\right] \vee \forall t \in \left[t_0, \hat{W}\right] \tag{5.3}$$

- achieving the above objective at minimum costs of transfers among reservoirs,
- achieving the above objective at minimum costs of water transport from reservoirs to WTP,
- reaching the above partial objectives while satisfying Eqs. (5.1) and (5.2).

W. Z. Chmielowski, *Management of Complex Multi-reservoir Water Distribution Systems Using Advanced Control Theoretic Tools and Techniques*, SpringerBriefs in Computational Intelligence, DOI: 10.1007/978-3-319-00239-2_5, © The Author(s) 2013

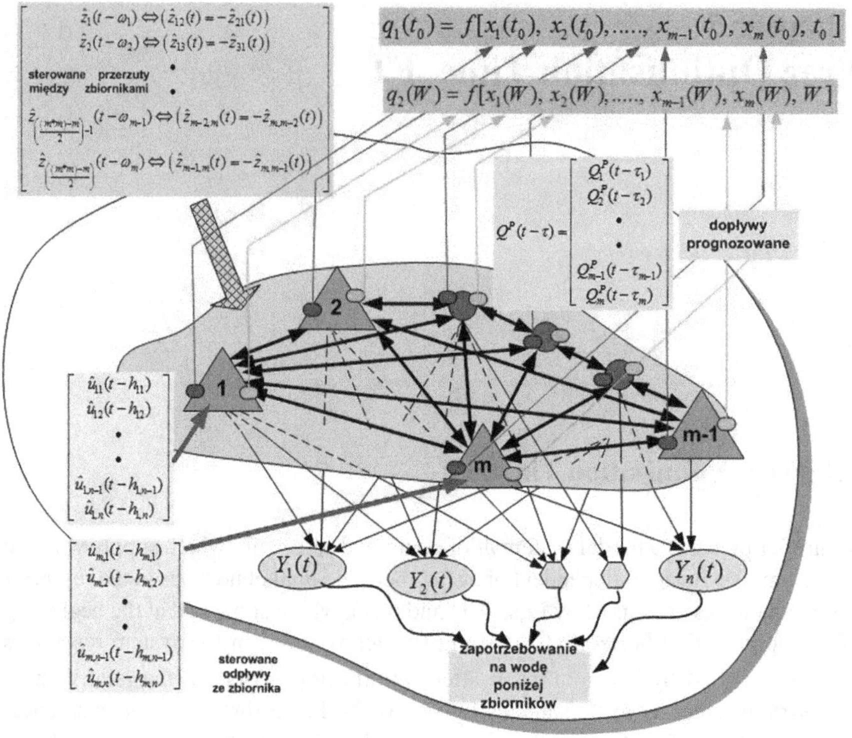

Fig. 5.1 The complex water-management system

The reservoirs supply water to n consumers (WTPs) at the same time, therefore further system description requires the introduction of a function for reservoir involvement in conducting the water demand function $Y_j(t),\ j = 1, \ldots, n\ \forall t \in [t_0^*, W] \vee \forall t \in [t_0, \hat{W}]$ dividing (for each time instant) the function $Y_j(t),\ j = 1, \ldots, n$ among the system reservoirs:

$$Y_j(t) - \sum_{i=1}^{m} b_{i,j}(t) \cdot Y_j = 0,\ j = 1, \ldots, n \tag{5.4}$$

In this reservoir system functioning (Fig. 5.1), there are a number of time delays with reference to individual system input and output variables. The above-mentioned delays, which significantly complicate the formal notation of the system function, reduce to the following dependencies:

- vector of delays related to water flow through the reservoirs

$$\boldsymbol{\tau}(t)^{\mathrm{T}} = \left[\, \tau_1(t) \bullet \tau_m(t) \,\right] \tag{5.5}$$

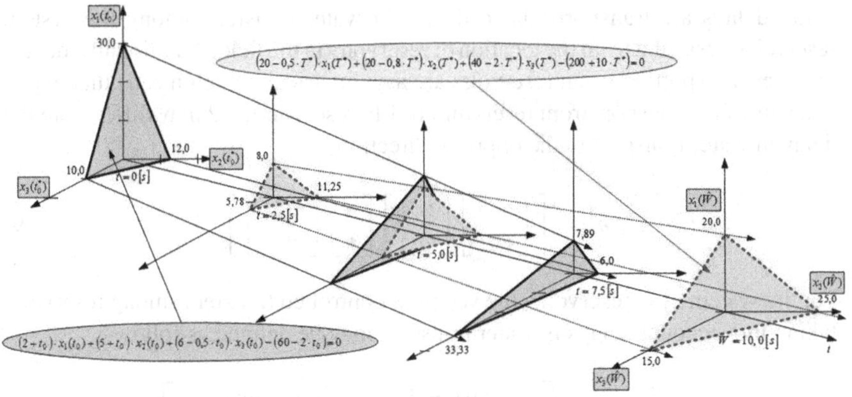

Fig. 5.2 Condition (4.1), (4.2) in space R3

In consequence, the vector of system reservoir inflows, taking into account the above-mentioned delays, should be defined as follows:

$$\boldsymbol{Q}^P(t-\boldsymbol{\tau}(t)) = \begin{bmatrix} Q_1^P(t-\tau_1(t)) \\ \bullet \\ Q_m^P(t-\tau_m(t)) \end{bmatrix} \tag{5.6}$$

- Transport-related delays in water distribution from the reservoir system to the WTP (5.7); the vector of controlled reservoir outflow taking into account the above-mentioned delays is defined by formula (5.8):

$$\boldsymbol{h}(t) = \begin{bmatrix} h_{11}(t) & \bullet & h_{m1}(t) \\ h_{12}(t) & \bullet & h_{m2}(t) \\ \bullet & \bullet & \bullet \\ h_{1n}(t) & \bullet & h_{mn}(t) \end{bmatrix} \tag{5.7}$$

$$\hat{\boldsymbol{u}}(t-\boldsymbol{h}(t))_{(+)} = \begin{bmatrix} \begin{bmatrix} u_{11}(t-h_{11}(t)) \\ u_{12}(t-h_{12}(t)) \\ \bullet \\ u_{1n}(t-h_{1n}(t)) \end{bmatrix} \\ \begin{bmatrix} \bullet \\ \bullet \\ \bullet \\ \bullet \end{bmatrix} \\ \begin{bmatrix} u_{m1}(t-h_{m1}(t)) \\ u_{m2}(t-h_{m2}(t)) \\ \bullet \\ u_{mn}(t-h_{mn}(t)) \end{bmatrix} \end{bmatrix} \tag{5.8}$$

- Other delays are transport-related delays in water transfers among the system reservoirs, resulting from the location of reservoirs in the field. Additionally, delays in water transport between reservoirs are asymmetrical, which means that, e.g. a delay in water transport from reservoir no. 1 to reservoir no. 2 may differ from the delay in water transport in the opposite direction.

$$\boldsymbol{\omega}^{\mathrm{T}} = \left[\begin{bmatrix} \omega_{12} \\ \omega_{21} \end{bmatrix} \begin{bmatrix} \omega_{23} \\ \omega_{32} \end{bmatrix} \begin{bmatrix} \bullet \\ \bullet \end{bmatrix} \begin{bmatrix} \omega_{m1} \\ \omega_{1m} \end{bmatrix} \right] \tag{5.9}$$

For the system of m reservoirs, the vector of controlled transfers among reservoirs taking into account delays in water transport may be defined as follows:

$$\hat{z}(t-\boldsymbol{\omega}) = \begin{bmatrix} \begin{bmatrix} z_{12}(t-\omega_{12}) \vee z_{21}(t-\omega_{21}) \\ z_{13}(t-\omega_{13}) \vee z_{31}(t-\omega_{31}) \\ z_{14}(t-\omega_{14}) \vee z_{41}(t-\omega_{41}) \\ \bullet \\ \bullet \\ \bullet \\ z_{1m}(t-\omega_{13}) \vee z_{m1}(t-\omega_{m1}) \end{bmatrix} \\ \begin{bmatrix} z_{23}(t-\omega_{23}) \vee z_{32}(t-\omega_{32}) \\ z_{24}(t-\omega_{24}) \vee z_{42}(t-\omega_{42}) \\ \bullet \\ \bullet \\ z_{2m}(t-\omega_{2m}) \vee z_{m2}(t-\omega_{m2}) \end{bmatrix} \\ \begin{bmatrix} z_{34}(t-\omega_{34}) \vee z_{43}(t-\omega_{43}) \\ z_{35}(t-\omega_{35}) \vee z_{53}(t-\omega_{53}) \\ \bullet \\ z_{3m}(t-\omega_{3m}) \vee z_{m3}(t-\omega_{m3}) \end{bmatrix} \\ \bullet \\ \begin{bmatrix} z_{m-2,m-1}(t-\omega_{m-2,m-1}) \\ \vee \\ z_{m-1,m-2}(t-\omega_{m-1,m-2}) \\ z_{m-2,m}(t-\omega_{m-2,m}) \\ \vee \\ z_{m,m-2}(t-\omega_{m,m-2}) \end{bmatrix} \\ \begin{bmatrix} z_{m-1,m}(t-\omega_{m-1,m}) \\ \vee \\ z_{m,-1m}(t-\omega_{m,m-1}) \end{bmatrix} \end{bmatrix} \tag{5.10}$$

5.1.1 Quality Coefficient

Optimal reservoir state trajectories and trajectories of outflow from reservoirs (control) will be determined by solving the following dynamic optimisation task, which consists of the following required elements:

$$F = 0,5 \cdot \int_{t_0^*, t_0}^{W, \hat{W}} \left\{ \begin{array}{l} [\mathbf{B}(t) \cdot \mathbf{Y}(t) \cdot \mathbf{S} \cdot \mathbf{1} - \boldsymbol{u}(t)]_{(+)}^{\mathrm{T}} \cdot \mathbf{A}_1 \cdot \\ \cdot [\mathbf{B}(t) \cdot \mathbf{Y}(t) \cdot \mathbf{S} \cdot \mathbf{1} - \boldsymbol{u}(t)]_{(+)} + \\ + \left[z(t)^{\mathrm{T}} \cdot \mathbf{A}_2 \cdot z(t) \right] + \\ + [\boldsymbol{u}(t)^{\mathrm{T}} \cdot \mathbf{A}_3 \cdot \boldsymbol{u}(t)] \end{array} \right\} dt \tag{5.11}$$

The following symbols have been used in Eq. (5.11) with reference to Fig. 5.1:

- t_0 steady optimisation start time at transient time of its end $\hat{W}$ [s]
- W steady optimisation end time at transient time of its start t_0^* [s]
- Matrix $\mathbf{B(t)}$ is a diagonal block matrix with terms constituting diagonal matrices, which consist of elements that are functions of reservoir involvement $i = 1, \ldots, m$ in carrying out the water demand function $Y_j(t)$, $j = 1, \ldots, n$, [m^3/s].

$$\mathbf{B}(t) = \begin{bmatrix} \mathbf{B}_1(t) & * & * \\ * & \bullet & * \\ * & * & \mathbf{B}_m(t) \end{bmatrix} \tag{5.12}$$

where:

$$\mathbf{B}_1(t) = \begin{bmatrix} b_{11}(t) & 0 & 0 & 0 \\ 0 & b_{12}(t) & 0 & 0 \\ 0 & 0 & \bullet & 0 \\ 0 & 0 & 0 & b_{1n}(t) \end{bmatrix} \tag{5.13}$$

$$\mathbf{B}_m(\mathbf{t}) = \begin{bmatrix} b_{m1}(t) & 0 & 0 & 0 \\ 0 & b_{m2}(t) & 0 & 0 \\ 0 & 0 & \bullet & 0 \\ 0 & 0 & 0 & b_{mn}(t) \end{bmatrix} \tag{5.14}$$

$$* = \begin{bmatrix} 0 & 0 & \bullet & 0 \\ 0 & 0 & \bullet & 0 \\ \bullet & \bullet & \bullet & \bullet \\ 0 & 0 & \bullet & 0 \end{bmatrix}_{(n*n)} \tag{5.15}$$

Matrix $\mathbf{Y(t)}$ is a diagonal block matrix with terms constituting diagonal matrices which consist of elements that are demand functions applicable in the system $Y_j(t)$, $j = 1, \ldots, 4$, [m^3/s]

$$\mathbf{Y}(t) = \begin{bmatrix} \circ(t) & * & * \\ * & \circ(t) & * \\ * & * & \circ(t) \end{bmatrix} \tag{5.16}$$

where:

$$\circ\,(t) = \begin{bmatrix} Y_1(t) & 0 & 0 & 0 \\ 0 & Y_2(t) & 0 & 0 \\ 0 & 0 & \bullet & 0 \\ 0 & 0 & 0 & Y_n(t) \end{bmatrix} \tag{5.17}$$

- The control vector (vector of controlled outflows from reservoirs) (5.8) is a block vector consisting of vectors, and elements of these vectors are outflows from reservoirs $i = 1, \ldots, m$ to conurbation $j = 1, \ldots, n$
- Then, matrix $\mathbf{A_1}$ is a positively determined block matrix with terms in a diagonal constituting diagonal matrices as well, which consist of elements that are weight coefficients connected with appropriate control vector elements

$$\mathbf{A_1} = \begin{bmatrix} \bullet_1 & * & * & * \\ * & \bullet_2 & * & * \\ * & * & \bullet & * \\ * & * & * & \bullet_m \end{bmatrix} \tag{5.18}$$

where:

$$\bullet_1 = \begin{bmatrix} a_{11} & 0 & 0 & 0 \\ 0 & a_{12} & 0 & 0 \\ 0 & 0 & a_{1*} & 0 \\ 0 & 0 & 0 & a_{1n} \end{bmatrix} \tag{5.19}$$

$$\bullet_2 = \begin{bmatrix} a_{21} & 0 & 0 & 0 \\ 0 & a_{22} & 0 & 0 \\ 0 & 0 & a_{2*} & 0 \\ 0 & 0 & 0 & a_{2n} \end{bmatrix} \tag{5.20}$$

$$\bullet_m = \begin{bmatrix} a_{m1} & 0 & 0 & 0 \\ 0 & a_{m2} & 0 & 0 \\ 0 & 0 & a_{m*} & 0 \\ 0 & 0 & 0 & a_{mn} \end{bmatrix} \tag{5.21}$$

- Matrix $\mathbf{S}$ is a structural matrix formed as a result of operation $\mathbf{S} = ((\mathbf{S_1}^{\mathrm{T}} \cdot \mathbf{S_1}) * \mathbf{I})$ (character $*$, table multiplication), where $\mathbf{S_1}$ according to (5.34), (5.35), and (5.36).

Later, a positively determined diagonal matrix $\mathbf{A_2}$ appears in a very interesting form,

$$\mathbf{A_2} = \begin{bmatrix} \mathbf{A}_{21} & 0_{(m-1,m-1)} & 0 & 0_{(2x2)} & 0 \\ 0_{(m\,x\,m)} & \mathbf{A}_{22} & 0 & 0_{(2x2)} & 0 \\ 0_{(m\,x\,m)} & 0_{(m-1,m-1)} & \bullet & 0_{(2x2)} & 0 \\ 0_{(m\,x\,m)} & 0_{(m-1,m-1)} & 0 & \mathbf{A}_{2,m-1} & 0 \\ 0_{(m\,x\,m)} & 0_{(m-1,m-1)} & 0 & 0_{(2x2)} & \mathbf{A}_{2,m} \end{bmatrix} \tag{5.22}$$

which consists of weight coefficients connected with a subintegral part of quality coefficient (5.11), corresponding to the costs of water transfers among reservoirs. The matrix dimensions match the dimensions of the reservoir transfer vector of (5.10), e.g $\mathbf{A_{21}}$ concerns the cost of transfer $z_{1,i}, (t), i = 1, \ldots, m$, that is from reservoir 1 to other reservoirs in the system, etc.

$$\mathbf{A_{21}} = \begin{bmatrix} a_{12} & & & \\ & \bullet & & \\ & & a_{1,m-1} & \\ & & & a_{1,m} \end{bmatrix}_{(m,m)} \tag{5.23}$$

$$\mathbf{A_{22}} = \begin{bmatrix} a_{23} & & & \\ & \bullet & & \\ & & a_{2,m-1} & \\ & & & a_{2,m} \end{bmatrix}_{(m-1,m-1)} \tag{5.24}$$

$$\mathbf{A_{2,m-1}} = \begin{bmatrix} a_{m-2,m-1} & \\ & a_{m-2,m} \end{bmatrix}_{(2,2)} \tag{5.25}$$

$$\mathbf{A_{2,m}} = \left[\mathbf{a_{m-1,m}}\right]_{(\mathbf{1,1})} \tag{5.26}$$

Structural unit vector:

$$\boldsymbol{1}^{\mathrm{T}} = [* \bullet *]_{1,m}, \quad * = [1\ 1 \bullet 1]_{1,n} \tag{5.27}$$

Considering the costs of water transport from reservoirs to WTP which appear in quality coefficient, it is necessary to introduce a matrix consisting of transport cost coefficients. Matrix $\mathbf{A_3}$ is a diagonal block matrix, which consists of elements that are costs of water transport from reservoirs $i = 1, \ldots, m$ do WTP $j = 1, \ldots, n$

$$\mathbf{A_3} = \begin{bmatrix} \mathbf{A_{31}} & * & * \\ * & \bullet & * \\ * & * & \mathbf{A_{3m}} \end{bmatrix} \tag{5.28}$$

$$\mathbf{A_{31}} = \begin{bmatrix} a_{11} & 0 & 0 & 0 \\ 0 & a_{12} & 0 & 0 \\ 0 & 0 & \bullet & 0 \\ 0 & 0 & 0 & a_{1n} \end{bmatrix} \tag{5.29}$$

$$\mathbf{A_{3m}} = \begin{bmatrix} a_{m1} & 0 & 0 & 0 \\ 0 & a_{m2} & 0 & 0 \\ 0 & 0 & \bullet & 0 \\ 0 & 0 & 0 & a_{mn} \end{bmatrix} \tag{5.30}$$

$$* = \begin{bmatrix} 0 & 0 & \bullet & 0 \\ 0 & 0 & \bullet & 0 \\ \bullet & \bullet & \bullet & \bullet \\ 0 & 0 & \bullet & 0 \end{bmatrix}_{(n*n)} \tag{5.31}$$

5.1.2 State Equation for Reservoirs

We have assumed a balance state equation for the system reservoirs

$$f : \dot{\boldsymbol{x}}(t) = \boldsymbol{Q}^P(t - \boldsymbol{\tau}(t)) - \mathbf{S_1}\boldsymbol{u}(t - \boldsymbol{h}(t)) + \mathbf{S_2}\boldsymbol{z}(t - \boldsymbol{\omega}) \tag{5.32}$$

The following symbols are used in the state Eq. (5.32):

- vector of reservoir state derivatives

$$\dot{\boldsymbol{x}}^{\mathrm{T}}(t) = [dx_1(t)/dt \;\; dx_2(t)/dt \;\; dx_3(t)/dt] \tag{5.33}$$

- vector of predicted inflows to reservoirs (5.6) $\boldsymbol{Q}^P(t - \boldsymbol{\tau}(t))$ $\left[\mathrm{m}^3/\mathrm{s}\right]$
- $\mathbf{S}_{1\,(\mathrm{diag}(m*m)_{(1*n)})}$, a diagonal block matrix needed to note the system state equation. Its terms in the main diagonal are vectors with elements with the values of either 0 or 1. The elements transmit information concerning the existence of connections between conurbations and reservoirs.
  If:

K—a set of indexes of possible connections between reservoir i, $i = 1, \ldots, m$ and WTPs j, $j = 1, \ldots, n$,
U—a set of indexes of ensuing connections between reservoir i and WTPs j, then:

$$\mathbf{S}_1 : \left\{i = 1, \ldots, m \;;\; s_{i,i} = *_{(1*n)}\right\} : \tag{5.34}$$

$$*_{(1*n)} : \left\{ \begin{array}{l} k = 1, \ldots, n \in K, \quad U \subset K \\ a_{1,k\in U} \quad = 1 \\ a_{1,k\in K\setminus U} = 0 \end{array} \right\} \tag{5.35}$$

$$\mathbf{S}_1 = \begin{bmatrix} * & [0] & [0] \\ [0] & \bullet & [0] \\ [0] & [0] & * \end{bmatrix} \quad [0] = \left[0\;0 \bullet 0\right] \tag{5.36}$$

Matrix $\mathbf{S_1}$ for the system from Fig. 5.3 is shown below, and later for the system from Fig. 5.4:

$$\mathbf{S_1} = \begin{bmatrix} [1\,1] & 0 & 0 & 0 \\ 0 & [1\,0] & 0 & 0 \\ 0 & 0 & [1\,1] & 0 \\ 0 & 0 & 0 & [0\,1] \end{bmatrix} \tag{5.37}$$

$$\mathbf{S_1} = \begin{bmatrix} [1\,1\,1\,0] & 0 \\ 0 & [0\,1\,1\,1] \end{bmatrix} \tag{5.38}$$

- $\boldsymbol{u}(t - \boldsymbol{h}(t))_{(+)}$ vector of controlled outflows from reservoirs to WTP,
- $\mathbf{z}(t - \boldsymbol{\omega})$ vector of controlled transfers among reservoirs,
- $\mathbf{S_2}$ structural matrix is needed to enter a link among reservoirs with reference to transfers among reservoirs in the system state equation. As regards the structure of vector of transfers among reservoirs (5.10), it is a little bothersome to write the structural matrix $\mathbf{S_2}$ which will be shown using a system of six reservoirs, later to go to a generalisation for m reservoirs. Without taking into account inflows $\boldsymbol{Q}^P(t - \boldsymbol{\tau}(t))$ to reservoirs, or either controlled outflows to WTP $\boldsymbol{u}(t - \boldsymbol{h}(t))_{(+)}$,

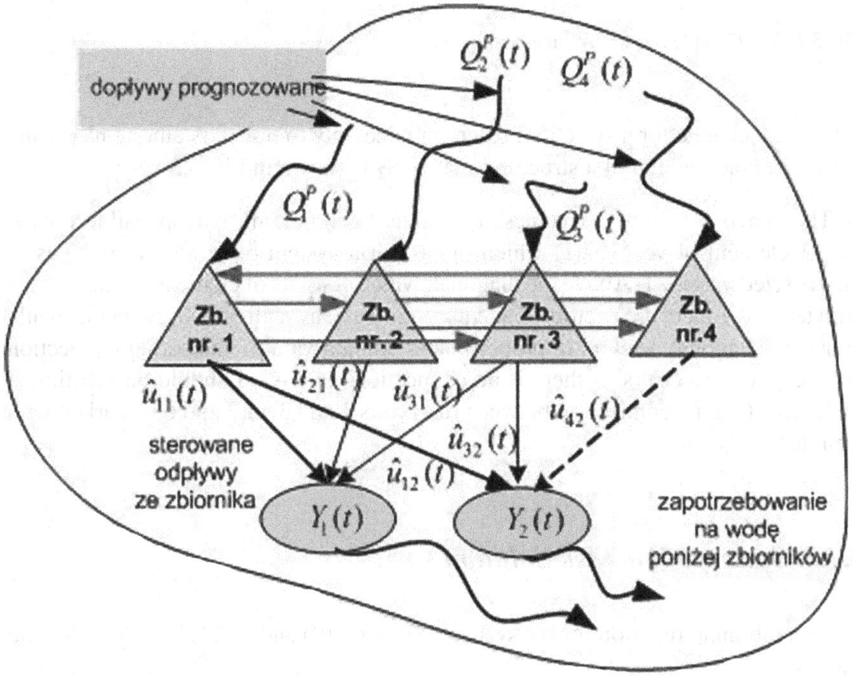

Fig. 5.3 The complex water-management system

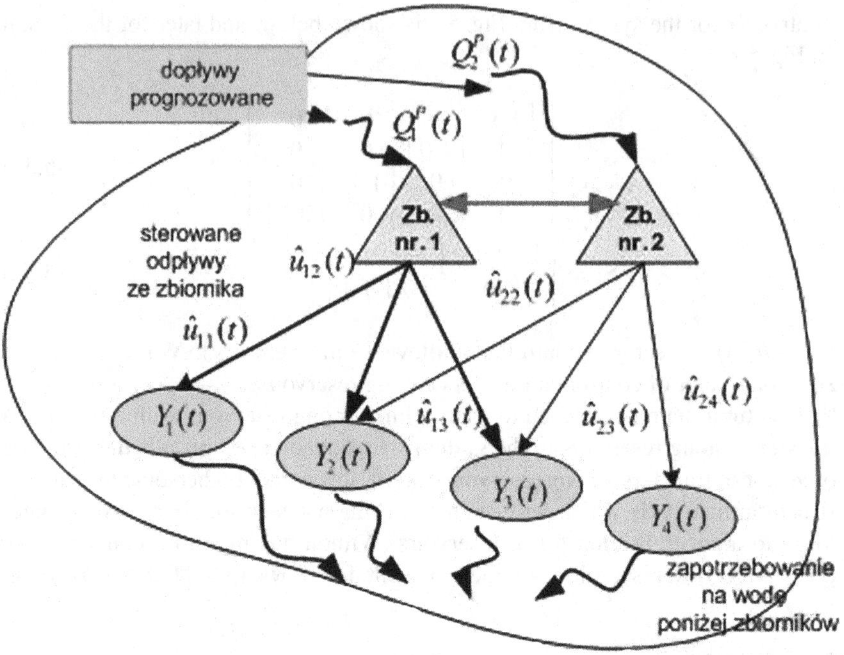

Fig. 5.4 The Complex water-management system

the state change for a system of reservoirs due only to transfers among reservoirs and the construction of a structural matrix $\mathbf{S_2}$ is shown in Fig. 5.5.

The matrix structure correctness is visible. Each column is responsible for the proper element of vector $\mathbf{z(t)}$ which in case of a system of m reservoirs leads to matrix sized $m, m-1$. Above the diagonal, which consists of elements being vectors with terms -1 there are vectors of adequate dimensions with only 0, elements, while under the diagonal, vectors in proper places contain value 1 indicating connection among given reservoirs. If there is no connection, pair -1, 1 should be substituted with zeros (e.g. no connection between reservoirs 1 and 3, or 3 and 6 is marked with a circle).

5.1.3 Optimisation Task Solution

The Hamiltonian function for the system of Eqs. (5.10) and (5.24) has the following form:

$$H = -f_0 + \boldsymbol{\psi}^{\mathrm{T}} \cdot f \tag{5.39}$$

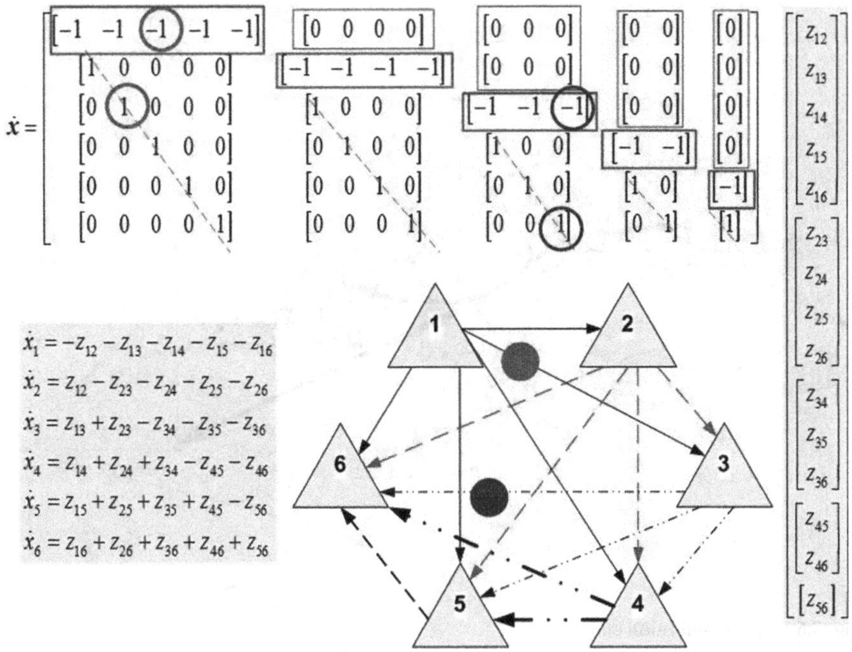

Fig. 5.5 Notation of matrix $\mathbf{S_2}$ for a system of six reservoirs

f_0—subintegral function of index (5.11)
$\boldsymbol{f}$—state equation for reservoirs (5.32)
$\boldsymbol{\psi}$—conjugate variable, vector [3*1]

$$H = -0,5 \cdot \left\{ \begin{array}{l} [\mathbf{B}(t) \cdot \mathbf{Y}(t) \cdot \mathbf{S} \cdot \boldsymbol{1} - \boldsymbol{u}(t)]^{\mathrm{T}}_{(+)} \cdot \mathbf{A}_1 \cdot \\ \cdot [\mathbf{B}(t) \cdot \mathbf{Y}(t) \cdot \mathbf{S} \cdot \boldsymbol{1} - \boldsymbol{u}(t)]_{(+)} + \\ + \left[z(t)^{\mathrm{T}} \cdot \mathbf{A}_2 \cdot z(t)\right] + \\ + [\boldsymbol{u}(t)^{\mathrm{T}} \cdot \ \mathbf{A}_3 \cdot \boldsymbol{u}(t)] \end{array} \right\} dt$$
$$+ \boldsymbol{\psi}(t)^{\mathrm{T}} \cdot \left[\boldsymbol{Q}^P(t - \boldsymbol{\tau}(t)) - \mathbf{S_1}\boldsymbol{u}(t - \boldsymbol{h}(t)) + \mathbf{S_2}z(t - \boldsymbol{\omega})\right] \tag{5.40}$$

System of equations for Hamiltonian function in form (5.40) is shown in points below:

1.

$$\left[(\nabla_u H)_{\hat{u},\hat{x},\hat{\psi}}\right]^{\mathrm{T}} = \boldsymbol{0} \Rightarrow \hat{\boldsymbol{u}}(t - \boldsymbol{h}(t))_{(+)} =$$
$$(\mathbf{A}_1 + \mathbf{A}_3)^{-1} \cdot \left[\mathbf{A}_1^{-1} \cdot \mathbf{B}(t) \cdot \mathbf{Y}(t) \cdot \mathbf{S} \cdot \boldsymbol{1} - \cdot \mathbf{S}_1{}^{\mathrm{T}} \cdot \hat{\boldsymbol{\psi}}(t)\right] \tag{5.41}$$

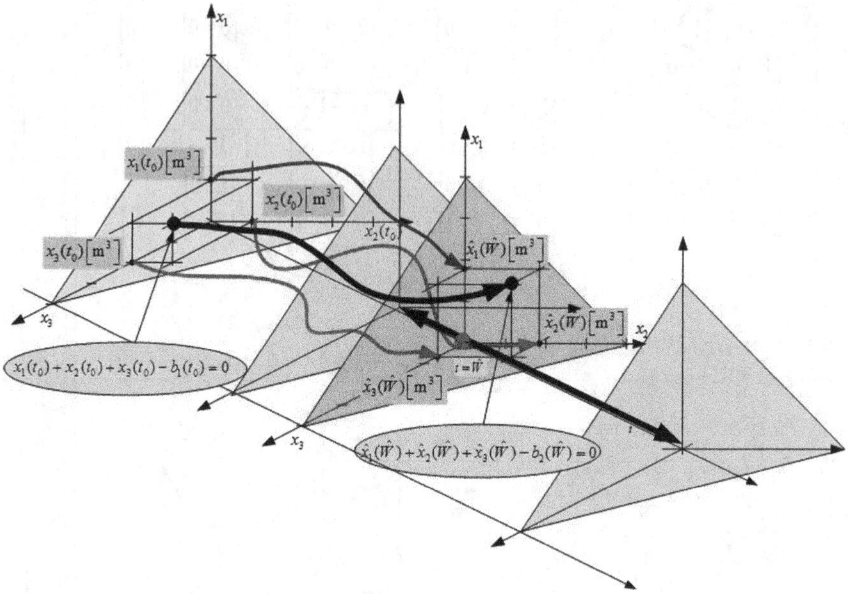

Fig. 5.6 Free optimisation end time

2.

$$\left[(\nabla_z H)_{\hat{u},\hat{z},\hat{x},\hat{\psi}}\right]^{\mathrm{T}} = \boldsymbol{0} \quad \Rightarrow \quad \hat{z}(t-\boldsymbol{\omega}) = \mathbf{A}_2^{-1} \cdot \mathbf{S}_2^{\mathrm{T}} \cdot \hat{\boldsymbol{\psi}}(t) \tag{5.42}$$

3.

$$\left[(\nabla_\psi H)_{\hat{u},\hat{x}}\right]^{\mathrm{T}} = \dot{\boldsymbol{x}}(t) \quad \Rightarrow \quad \boldsymbol{Q}^P(t-\boldsymbol{\tau}(t)) - \mathbf{S}_1 u(\mathbf{t}-h(\mathbf{t})) + \mathbf{S}_2 z(\mathbf{t}-\omega) \tag{5.43}$$

4.

$$\left[-(\nabla_x H)_{\hat{u},\hat{x},\hat{\psi}}\right]^{\mathrm{T}} = \dot{\hat{\boldsymbol{\psi}}}(t) \Rightarrow \dot{\hat{\boldsymbol{\psi}}}(t) = \boldsymbol{0}_{(3*1)} \tag{5.44}$$

Equation (5.44) shows that:

$$\hat{\boldsymbol{\psi}}(t) = \boldsymbol{C} \tag{5.45}$$

5.1.3.1 Steady Start Time t_0 Transient End Time $\hat{W}$

If end condition (4.2) is a function of the system state and time vector, in general its variation has the following form:

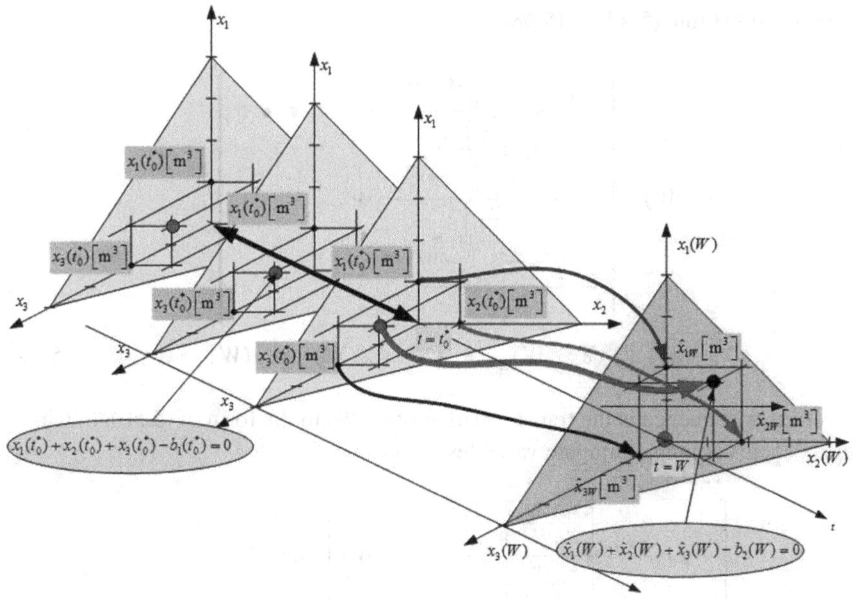

Fig. 5.7 Free optimisation start time

$$\left[\frac{\partial h_2(W)}{\partial x_1(t)}\right]_{t=W} \cdot \delta x_1(W) + \left[\frac{\partial h_2(W)}{\partial x_2(t)}\right]_{t=W} \cdot \delta x_2(W) + \bullet$$
$$\left[\frac{\partial h_2(W)}{\partial x_m(t)}\right]_{t=W} \cdot \delta x_m(W) + \left[\frac{\partial h_2(W)}{\partial t}\right]_{t=W} \cdot \delta t_0 = 0 \tag{5.46}$$

On the basis of the previous equation, the differential $\delta x_1(W)$ may be written as follows:

$$\delta x_1(W) = -\frac{\left[\frac{\partial h_2(W)}{\partial x_2(t)}\right]_{t=W} \cdot \delta x_2(W)}{\left[\frac{\partial h_2(W)}{\partial x_1(t)}\right]_{t=W}} \bullet \bullet$$
$$-\frac{\left[\frac{\partial h_2(W)}{\partial x_m(t)}\right]_{t=W} \cdot \delta x_m(W)}{\left[\frac{\partial h_2(W)}{\partial x_1(t)}\right]_{t=W}} - \frac{\left[\frac{\partial h_2(W)}{\partial t}\right]_{t=W} \cdot \delta W}{\left[\frac{\partial h_2(W)}{\partial x_1(t)}\right]_{t=W}} \tag{5.47}$$

Using (5.47), the transversality condition of right optimisation interval end may be written as follows:

$$\psi_1(W) \cdot \delta x_1(W) + \psi_2(W) \cdot \delta x_2(W) + \bullet \bullet$$
$$+ \psi_m(W) \cdot \delta x_m(W) = 0 \tag{5.48}$$

Then, we substitute (5.47) in (5.48)

$$\psi_1(W)\cdot\left\{\begin{array}{l} -\dfrac{\left[\frac{\partial h_2(W)}{\partial x_2(t)}\right]_{t=W}\cdot\delta x_2(W)}{\left[\frac{\partial h_2(W)}{\partial x_1(t)}\right]_{t=W}} \quad - \quad \bullet\ \bullet + \\ -\dfrac{\left[\frac{\partial h_2(W)}{\partial x_m(t)}\right]_{t=W}\cdot\delta x_m(W)}{\left[\frac{\partial h_2(W)}{\partial x_1(t)}\right]_{t=W}} \quad + \\ -\dfrac{\left[\frac{\partial h_2(W)}{\partial t}\right]_{t=W}\cdot\delta W}{\left[\frac{\partial h_2(W)}{\partial x_1(t)}\right]_{t=W}} \end{array}\right\}$$
$$+\,\psi_2(W)\cdot\delta x_2(W)+\bullet\ \bullet+\psi_m(W)\cdot\delta x_m(W)=0 \tag{5.49}$$

A further sequence is the transformation of (5.49) to the form of (5.50), in which dependencies among conjugate variables are visible

$$\begin{aligned} &\left[-\psi_1(W)\cdot\frac{\left[\frac{\partial h_2(W)}{\partial x_2(t)}\right]_{t=W}}{\left[\frac{\partial h_2(W)}{\partial x_1(t)}\right]_{t=W}}+\psi_2(W)\right]\cdot\delta x_2(W)\\ &+\left[-\psi_1(W)\cdot\frac{\left[\frac{\partial h_2(W)}{\partial x_3(t)}\right]_{t=W}}{\left[\frac{\partial h_2(W)}{\partial x_1(t)}\right]_{t=W}}+\psi_3(W)\right]\cdot\delta x_3(W)\\ &+\bullet\quad\left[-\psi_1(W)\cdot\frac{\left[\frac{\partial h_2(W)}{\partial x_m(t)}\right]_{t=W}}{\left[\frac{\partial h_2(W)}{\partial x_1(t)}\right]_{t=W}}+\psi_m(W)\right]\cdot\delta x_m(W)\\ &+\left[-\psi_1(W)\cdot\frac{\left[\frac{\partial h_2(W)}{\partial t}\right]_{t=W}\cdot\delta W}{\left[\frac{\partial h_2(W)}{\partial x_1(t)}\right]_{t=W}}\right]=0 \end{aligned} \tag{5.50}$$

As differentials $\delta x_2(W), \delta x_3(W), \ldots, \delta x_m(W), \delta W$ may have different signs and $\psi_1(W) \neq 0$ taking into account (5.45) written out as elements $\psi_1(W) = C_1$, $\psi 2(W) = C_2, \ldots, \psi_m(W) = C_m$ the following relationships can be seen:

$$C_2=C_1\cdot\frac{\left[\frac{\partial h_2(W)}{\partial x_2(t)}\right]_{t=W}}{\left[\frac{\partial h_2(W)}{\partial x_1(t)}\right]_{t=W}}\quad C_3=C_1\cdot\frac{\left[\frac{\partial h_2(W)}{\partial x_3(t)}\right]_{t=W}}{\left[\frac{\partial h_2(W)}{\partial x_1(t)}\right]_{t=W}}$$
$$\bullet\ \bullet\qquad C_m=C_1\cdot\frac{\left[\frac{\partial h_2(W)}{\partial x_m(t)}\right]_{t=W}}{\left[\frac{\partial h_2(W)}{\partial x_1(t)}\right]_{t=W}} \tag{5.51}$$

$$C_1 \cdot \frac{\left[\frac{\partial h_2(W)}{\partial t}\right]_{t=W}}{\left[\frac{\partial h_2(W)}{\partial x_1(t)}\right]_{t=W}} = 0 \tag{5.52}$$

We write Eq. (5.45) in a new form

$$\boldsymbol{\psi}(W) = \boldsymbol{C} = \begin{bmatrix} 1 \\ \frac{\left[\frac{\partial h_2(W)}{\partial x_2(t)}\right]_{t=W}}{\left[\frac{\partial h_2(W)}{\partial x_1(t)}\right]_{t=W}} \\ \frac{\left[\frac{\partial h_2(W)}{\partial x_3(t)}\right]_{t=W}}{\left[\frac{\partial h_2(W)}{\partial x_1(t)}\right]_{t=W}} \\ \bullet \\ \frac{\left[\frac{\partial h_2(W)}{\partial x_m(t)}\right]_{t=W}}{\left[\frac{\partial h_2(W)}{\partial x_1(t)}\right]_{t=W}} \end{bmatrix} \cdot C_1 = \boldsymbol{F} \cdot C_1 \tag{5.53}$$

Now, Eqs. (5.41) and (5.42) have the following form:

$$\begin{aligned} \hat{\boldsymbol{u}}(t - \boldsymbol{h}(t))_{(+)} &= (\mathbf{A_1} + \mathbf{A_3})^{-1} \\ &\cdot \left[\mathbf{A}_1^{-1} \cdot \mathbf{B}(t) \cdot \mathbf{Y}(t) \cdot \mathbf{S} \cdot \boldsymbol{1} - \cdot \mathbf{S}_1{}^{\mathrm{T}} \cdot \boldsymbol{F} \cdot \hat{C}_1\right] \end{aligned} \tag{5.54}$$

$$\hat{\boldsymbol{z}}(t - \boldsymbol{\omega}) = \mathbf{A}_2^{-1} \cdot \mathbf{S}_2^{\mathrm{T}} \cdot \boldsymbol{F} \cdot \hat{C}_1 \tag{5.55}$$

Using Eq. (5.43), we may write the equation of the system reservoir state trajectories in the following form:

$$\boldsymbol{x}(t) = \boldsymbol{x}_0(t_0) + \int_{t_0}^{t} \begin{bmatrix} \boldsymbol{Q}^P(\xi - \boldsymbol{\tau}(\xi)) - \\ +\mathbf{S}_1 \cdot \hat{\boldsymbol{u}}((\xi) - \boldsymbol{h}(\xi)) + \\ +\mathbf{S}_2 \cdot \hat{\boldsymbol{z}}(\xi - \omega) \end{bmatrix} d\xi \tag{5.56}$$

Further, we will aim at determining constant C_1. In Eq. (5.56) we substitute (5.54) and (5.55) and assume an upper integration boundary equal to an unknown end optimisation time.

$$\boldsymbol{x}(\hat{W}) = \boldsymbol{x}_0(t_0) + \int_{t_0}^{\hat{W}} \begin{bmatrix} \boldsymbol{Q}^P(t - \boldsymbol{\tau}(t)) - \\ +\mathbf{S}_1\,(\mathbf{A_1} + \mathbf{A_3})^{-1} \\ \cdot\,[\,\mathbf{A}_1^{-1} \cdot \mathbf{B}(t) \cdot \mathbf{Y}(t) \cdot \mathbf{S} \cdot \boldsymbol{1} \\ - \cdot \mathbf{S}_1^{\mathrm{T}} \cdot \boldsymbol{F} \cdot \hat{C}_1\,] + \\ +\mathbf{S_2} \cdot \mathbf{A}_2^{-1} \cdot \mathbf{S}_2^{\mathrm{T}} \cdot \boldsymbol{F} \cdot \hat{C}_1 \end{bmatrix} dt \tag{5.57}$$

We rearrange Eq. (5.57) integrating expressions constant in time

$$
\begin{aligned}
\boldsymbol{x}(\hat{W}) = \boldsymbol{x}_0(t_0) + \\
+ \int_{t_0}^{\hat{W}} \begin{bmatrix} \boldsymbol{Q}^P(t-\tau(t)) - \\ +\mathbf{S}_1\,(\mathbf{A}_1+\mathbf{A}_3)^{-1} \cdot \\ \cdot\,[\,\mathbf{A}_1^{-1}\cdot\mathbf{B}(t)\cdot\mathbf{Y}(t)\cdot\mathbf{S}\cdot\boldsymbol{1} \end{bmatrix} dt \\
+ \left[\mathbf{S}_1\,(\mathbf{A}_1+\mathbf{A}_3)^{-1}\cdot\mathbf{S}_1^{\mathrm{T}} + \mathbf{S}_2\cdot\mathbf{A}_2^{-1}\cdot\mathbf{S}_2^{\mathrm{T}}\right] \\
\cdot\,\boldsymbol{F}\cdot\hat{C}_1\cdot(\hat{W}-t_0)
\end{aligned}
\tag{5.58}
$$

Then, we substitute (5.58) in the equation of bounded final conditions (4.2)

$$
\left[e_1(\hat{W}) \bullet e_m(\hat{W})\right] \cdot \begin{bmatrix} x_1(\hat{W}) \\ \bullet \\ x_m(\hat{W}) \end{bmatrix} - b_2(\hat{W}) = 0 \tag{5.59}
$$

$$
\begin{aligned}
\boldsymbol{E}(\hat{W}) \cdot \begin{Bmatrix} \boldsymbol{x}_0(t_0) + \\ \int_{t_0}^{\hat{W}} \begin{Bmatrix} \boldsymbol{Q}^P(t-\tau(t)) - \\ +\,\mathbf{S}_1\,(\mathbf{A}_1+\mathbf{A}_3)^{-1}\cdot \\ \cdot\mathbf{A}_1^{-1}\cdot\mathbf{B}(t)\cdot\mathbf{Y}(t)\cdot\mathbf{S}\cdot 1 \end{Bmatrix} dt \\ + \begin{Bmatrix} \mathbf{S}_1\,(\mathbf{A}_1+\mathbf{A}_3)^{-1}\cdot\mathbf{S}_1^T + \\ +\mathbf{S}_2\cdot\mathbf{A}_2\cdot\mathbf{S}_2^T \end{Bmatrix} \cdot \\ \cdot\boldsymbol{F}\cdot\hat{C}_1\cdot\left(\hat{W}-t_0\right) \end{Bmatrix} \\
- b_2(\hat{W}) = 0
\end{aligned}
\tag{5.60}
$$

After further transformations, we receive first relation specifying C_1 needed to determine optimisation end time $\hat{W}$

$$
\hat{C}_1 = \begin{Bmatrix} \boldsymbol{E}(\hat{W}) \cdot \begin{Bmatrix} \mathbf{S}_1\,(\mathbf{A}_1+\mathbf{A}_3)^{-1}\cdot\mathbf{S}_1^{\mathrm{T}} + \\ +\mathbf{S}_2\cdot\mathbf{A}_2\cdot\mathbf{S}_2^{\mathrm{T}} \end{Bmatrix} \cdot \\ \cdot\boldsymbol{F}\cdot\left(\hat{W}-t_0\right) \end{Bmatrix}^{-1} \times
\begin{Bmatrix} b_2(\hat{W}) - \\ \boldsymbol{E}(\hat{W})\cdot \\ \begin{Bmatrix} \int_{t_0}^{\hat{W}} \begin{Bmatrix} \boldsymbol{Q}^P(t-\tau(t)) - \\ +\,\mathbf{S}_1\,(\mathbf{A}_1+\mathbf{A}_3)^{-1}\cdot \\ \cdot\mathbf{A}_1^{-1}\cdot\mathbf{B}(t)\cdot\mathbf{Y}(t)\cdot\mathbf{S}\cdot 1 \end{Bmatrix} dt + \\ +\boldsymbol{x}_0(t_0) \end{Bmatrix} \end{Bmatrix}
\tag{5.61}
$$

In order to simplify formula (5.61), we will make the following substitutions:

$$\mathbf{P_1} = \mathbf{S_1}\,(\mathbf{A_1} + \mathbf{A_3})^{-1} \cdot \mathbf{S_1^T} + \mathbf{S_2} \cdot \mathbf{A_2} \cdot \mathbf{S_2^T} \tag{5.62}$$

$$\boldsymbol{P}_2 = \int_{t_0}^{\hat{W}} \left\{ \begin{array}{l} \boldsymbol{Q}^P(t - \tau(t)) + \\ + \mathbf{S}_1\,(\mathbf{A_1} + \mathbf{A_3})^{-1} \cdot \\ \cdot \mathbf{A}_1^{-1} \cdot \mathbf{B}(t) \cdot \mathbf{Y}(t) \cdot \mathbf{S} \cdot 1 \end{array} \right\} dt \tag{5.63}$$

After substituting (5.61) and (5.62) in (5.61) we receive simple relationship describing variable C_1

$$\hat{C}_1 = \left\{ \boldsymbol{E}(\hat{W}) \cdot \mathbf{P}_1 \cdot \boldsymbol{F} \cdot (\hat{W} - t_0) \right\}^{-1} \cdot \left\{ b_2(\hat{W}) - \boldsymbol{E}(\hat{W}) \cdot (\boldsymbol{P}_2 - \boldsymbol{x}_0(t_0)) \right\} \tag{5.64}$$

In order to compute time $\hat{W}$ for optimisation end, another equation is necessary, which we may obtain using the following relationship:

$$H(\hat{W}) - \left(\frac{\partial K\left[x(\hat{W}), \hat{W}\right]}{\partial \hat{W}} \right) = 0 \tag{5.65}$$

In the case under consideration, index (5.11) does not contain the function of final conditions $K[x(\hat{W}, \hat{W}]$, therefore Hamiltonian function $H(\hat{W}) = 0$. Later, we transform the Hamiltonian function aiming at determining $\boldsymbol{\psi}(\hat{W}) = \hat{C}_1$ We substitute relationships (5.67) and (5.68) in Eq. (5.66), and then, after extended transformations, we receive another dependency describing constant C_1 in function of the sought after optimisation end time $\hat{W}$ (5.69)

$$-0,5 \cdot \left\{ \begin{array}{l} [\mathbf{B}(t) \cdot \mathbf{Y}(t) \cdot \mathbf{S} \cdot \mathbf{1} - \boldsymbol{u}(t)]_{(+)}^{\mathrm{T}} \cdot \mathbf{A}_1 \cdot \\ \cdot [\mathbf{B}(t) \cdot \mathbf{Y}(t) \cdot \mathbf{S} \cdot \mathbf{1} - \boldsymbol{u}(t)]_{(+)} + \\ + \left[z(t)^{\mathrm{T}} \cdot \mathbf{A}_2 \cdot z(t)\right] + \\ + [\boldsymbol{u}(t)^{\mathrm{T}} \cdot \mathbf{A}_3 \cdot \boldsymbol{u}(t)] \end{array} \right\} + \boldsymbol{\psi}(t)^{\mathrm{T}} \cdot \left[\boldsymbol{Q}^P(t - \boldsymbol{\tau}(t)) - \mathbf{S}_1 u(\mathbf{t} - h(\mathbf{t})) + \mathbf{S}_2 z(\mathbf{t} - \omega)\right] = 0 \tag{5.66}$$

$$\hat{\boldsymbol{u}}(t - \boldsymbol{h}(t))_{(+)} = (\mathbf{A_1} + \mathbf{A_2})^{-1} \cdot \left[\mathbf{A}_1^{-1} \cdot \mathbf{B}(t) \cdot \mathbf{Y}(t) \cdot \mathbf{S} \cdot \boldsymbol{1} - \cdot \mathbf{S}_1{}^{\mathrm{T}} \cdot \boldsymbol{F} \cdot \hat{C}_1\right] \tag{5.67}$$

$$\hat{z}(t - \omega) = \mathbf{A}_2^{-1} \cdot \mathbf{S}_2^{\mathrm{T}} \cdot \boldsymbol{F} \cdot \hat{C}_1 \tag{5.68}$$

$$L_1(\hat{W}) \cdot \hat{C}_1^2 + \left(L_2(\hat{W}) + L_3(\hat{W})\right) \cdot \hat{C}_1 - 0.5 \cdot L_4 = 0 \tag{5.69}$$

Quadratic Eq. (5.69) with regard to C_1 contains a series of coefficients introduced during transformations of Eq. (5.66). They include:

$$\begin{aligned} L_1(\hat{W}) &= R_{12}(\hat{W}) - 0,5 \cdot R_7(\hat{W}) \quad (1x1) \\ L_2(\hat{W}) &= R_{11}(\hat{W}) - 0,5 \cdot R_8(\hat{W}) \quad (1x1) \\ L_3(\hat{W}) &= 0,5 \cdot R_9(\hat{W}) \quad (1x1) \\ L_4(\hat{W}) &= 0,5 \cdot R_{10}(\hat{W}) \quad (1x1) \end{aligned} \tag{5.70}$$

$$\begin{aligned} R_{12}(\hat{W}) &= \boldsymbol{F}^T(\hat{W}) \cdot (\mathbf{S_1} \cdot \boldsymbol{R}_3(\hat{W}) + \mathbf{S_2} \cdot \boldsymbol{R}_4(\hat{W})) \\ R_{11}(\hat{W}) &= \boldsymbol{F}^{\mathrm{T}}(\hat{W}) \cdot (\boldsymbol{Q}^P(\hat{W}) - \mathbf{S_1} \cdot \boldsymbol{R}_2(\hat{W})) \\ R_{10}(\hat{W}) &= \boldsymbol{R}_5(\hat{W})^{\mathrm{T}} \cdot \mathbf{A_1} \cdot \boldsymbol{R}_5(\hat{W}) + \\ &\quad + \boldsymbol{R}_2(\hat{W})^{\mathrm{T}} \cdot \mathbf{A_3} \cdot \boldsymbol{R}_2(\hat{W}) \\ R_9(\hat{W}) &= \boldsymbol{R}_5(\hat{W})^{\mathrm{T}} \cdot \mathbf{A_1} \cdot \boldsymbol{R}_3(\hat{W}) + \\ &\quad + \boldsymbol{R}_2(\hat{W})^{\mathrm{T}} \cdot \mathbf{A_3} \cdot \boldsymbol{R}_3(\hat{W}) \\ R_8(\hat{W}) &= \boldsymbol{R}_3(\hat{W})^{\mathrm{T}} \cdot \mathbf{A_1} \cdot \boldsymbol{R}_5(\hat{W}) + \\ &\quad + \boldsymbol{R}_3(\hat{W})^{\mathrm{T}} \cdot \mathbf{A_3} \cdot \boldsymbol{R}_2(\hat{W}) \\ R_7(\hat{W}) &= \boldsymbol{R}_3(\hat{W})^{\mathrm{T}} \cdot \mathbf{A_1} \cdot \boldsymbol{R}_3(\hat{W}) + R_6(\hat{W}) + \\ &\quad + \boldsymbol{R}_3(\hat{W})^{\mathrm{T}} \cdot \boldsymbol{R}_3(\hat{W}) \\ R_6(\hat{W}) &= \boldsymbol{R}_4(\hat{W})^{\mathrm{T}} \cdot \mathbf{A_2} \cdot \boldsymbol{R}_4(\hat{W}) \\ \boldsymbol{R}_5(\hat{W}) &= \boldsymbol{R}_1(\hat{W}) - \boldsymbol{R}_2(\hat{W}) \\ \boldsymbol{R}_4(\hat{W}) &= \mathbf{A_2}(\hat{W})^{-1} \cdot \mathbf{S_2}^T \cdot \boldsymbol{F} \\ \boldsymbol{R}_3(\hat{W}) &= (\mathbf{A_1} + \mathbf{A_2})^{-1} \cdot \mathbf{S_1}^T \cdot \boldsymbol{F} \\ \boldsymbol{R}_2(\hat{W}) &= (\mathbf{A_1} + \mathbf{A_3})^{-1} \cdot \mathbf{A_1} \cdot \boldsymbol{R}_1 \\ \boldsymbol{R}_1(\hat{W}) &= \mathbf{B}(\hat{W}) \cdot \mathbf{Y}(\hat{W}) \cdot \mathbf{S} \cdot 1 \end{aligned} \tag{5.71}$$

First, further solving of this task comes down to determining the optimal optimisation end time. $\hat{W}$ is determined by way of comparing Eqs. (5.64) and (5.69). An analytical solution to this problem is possible only for very simple functions describing the system input parameters, including predicted inflow, demand functions, boundary conditions, and others. Digital solution remains in cases of more complex descriptions of input parameters. Among other things, Eq. (5.64) is a function of the initial filling levels in reservoirs, which are linked by Eq. (4.1). Therefore, it is necessary to determine optimisation start time t_0, and then to choose any point satisfying Eq. (4.1) for steady t_0. The course of transformations and substitutions completed so far and planned in future indicates that the end solution will be a function of an assumed vector of initial filling levels in reservoirs $\boldsymbol{x}(\boldsymbol{t_0})^{\mathrm{T}} = [x_1(t_0), \bullet, x_m(t_0)]$. Any change in the vector of initial filling levels in reservoirs at other input parameters remaining unchanged gives a new solution to the whole task.

A further solution will be obtained without any major difficulties. We substitute $\hat{C}_1$ obtained from comparison of (5.64) and (5.69) in (5.41) and (5.42), thus receiving the optimal control vector $\hat{\boldsymbol{u}}(t-\boldsymbol{h}(t))_{(+)}$, $\forall t \in [t_0, W]$, and vector of transfers among reservoirs $\hat{\mathbf{z}}(t-\boldsymbol{\omega})$.

Transport-related delays in transfers among reservoirs have a fundamental effect on the implementation of the end condition of reservoir state trajectories (4.2). They force a change in the transfer vector value so as to satisfy condition (4.2). In order to satisfy the required condition (4.2) there is a need for a changed water distribution among reservoirs. This change results from the diagram shown below. Formula (5.41) specifies suitably increased/reduced transfer among reservoirs during distribution period, taking into account delays.

$$\mathbf{z}(t-\omega) = \begin{bmatrix} z_1(t-\omega_{12\vee 21}) : \otimes_1 \\ z_2(t-\omega_{23\vee 32}) : \otimes_2 \\ z_3(t-\omega_{31\vee 13}) : \otimes_3 \end{bmatrix}$$

$$\otimes_1 = \left\{ \begin{array}{c} |z| = |z_{12}| = |z_{21}| \\ (\mathrm{sgn}(z_1) = 1 \wedge \omega_{12} > 0) \Rightarrow \\ z_1(t-\omega_{12}) = |z_{12}| + \frac{\int_0^{\omega_{12}} |z_{12}| dt}{(W-\omega_{12})}; \\ (\mathrm{sgn}(z_1) = -1 \wedge \omega_{21} > 0) \Rightarrow \\ z_1(t-\omega_{21}) = |z_{12}| + \frac{\int_0^{\omega_{21}} |z_{12}| dt}{(W-\omega_{21})} \end{array} \right\} \tag{5.72}$$

$$\otimes_2 = \left\{ \begin{array}{c} |z| = |z_{23}| = |z_{32}| \\ (\mathrm{sgn}(z_2) = 1 \wedge \omega_{23} > 0) \Rightarrow \\ z_2(t-\omega_{23}) = |z_{23}| + \frac{\int_0^{\omega_{23}} |z_{23}| dt}{(W-\omega_{23})} \\ (\mathrm{sgn}(z_2) = -1 \wedge \omega_{32} > 0) \Rightarrow \\ z_2(t-\omega_{32}) = |z_{32}| + \frac{\int_0^{\omega_{32}} |z_{23}| dt}{(W-\omega_{32})} \end{array} \right\}$$

$$\otimes_3 = \left\{ \begin{array}{c} |z| = |z_{31}| = |z_{13}| \\ (\mathrm{sgn}(z_3) = 1 \wedge \omega_{31} > 0) \Rightarrow \\ z_3(t-\omega_{31}) = |z_{31}| + \frac{\int_0^{\omega_{31}} |z_{31}| dt}{(W-\omega_{31})} \\ (\mathrm{sgn}(z_3) = -1 \wedge \omega_{13} > 0) \Rightarrow \\ z_3(t-\omega_{13}) = |z_{13}| + \frac{\int_0^{\omega_{13}} |z_{13}| dt}{(W-\omega_{13})} \end{array} \right\} \tag{5.73}$$

Formulas (5.72) and (5.73) take into account the possibility of water transfer both ways (e.g. from reservoir no. 1 to reservoir no. 2 and the other way round) and asymmetrical transport-related delay (e.g. $\omega_{12} \neq \omega_{21}$). This principle is applicable to all links among reservoirs. With reference to our water system (Fig. 5.1), writing modifications in the vector of transfers $\hat{\mathbf{z}}(t-\omega)$ formula (5.72) and (5.73) is relatively simple. The degree of complexity of this notation increases very rapidly with the system dimensionality, and is in particular determined by the number and direction of transfers among reservoirs and related delays. Only after the implementation of the previously presented modifications can we substitute (5.54) and (5.57) in (5.58) to obtain the vector of optimal trajectories of states, satisfying condition (4.2). We will receive the minimal quality coefficient value when substituting (5.54) and (5.57) in (5.11).

5.1.3.2 Free Start Time t_0^*, Steady End Time W

This case, unlike the previous one, comes down to determining the optimal time for optimisation start t_0^*. If optimisation is started at this moment, the trajectories of the system reservoir states with initial conditions satisfying Eq. (4.1), will fulfil Eq. (4.2) at the end of optimisation horizon W binding these trajectories in the final instant. Methodically, mathematical notation is similar to that shown in previous chapter. If final condition (4.1) is a function of system state and time vector, in general its variation has the following form:

$$\left[\frac{\partial h_1\left(t_0^*\right)}{\partial x_1(t)}\right]_{t=t_0^*} \cdot \delta x_1(t_0^*) + \left[\frac{\partial h_1\left(t_0^*\right)}{\partial x_2(t)}\right]_{t=t_0^*} \cdot \delta x_2(t_0^*) + \bullet$$
$$\left[\frac{\partial h_1\left(t_0^*\right)}{\partial x_m(t)}\right]_{t=t_0^*} \cdot \delta x_m(t_0^*) + \left[\frac{\partial h_1\left(t_0^*\right)}{\partial t}\right]_{t=t_0^*} \cdot \delta t_0 = 0 \tag{5.74}$$

On the basis of the previous equation, the differential $\delta x_1(t_0^*)$ may be written as follows:

$$\delta x_1(W) = -\frac{\left[\frac{\partial h_1(t_0^*)}{\partial x_2(t)}\right]_{t=t_0^*} \cdot \delta x_2(t_0^*)}{\left[\frac{\partial h_1(t_0^*)}{\partial x_1(t)}\right]_{t=t_0^*}} + \bullet\bullet$$
$$-\frac{\left[\frac{\partial h_1(t_0^*)}{\partial x_m(t)}\right]_{t=t_0^*} \cdot \delta x_m(t_0^*)}{\left[\frac{\partial h_1(t_0^*)}{\partial x_1(t)}\right]_{t=t_0^*}} - \frac{\left[\frac{\partial h_1(t_0^*)}{\partial t}\right]_{t=t_0^*} \cdot \delta t_0^*}{\left[\frac{\partial h_1(t_0^*)}{\partial x_1(t)}\right]_{t=t_0^*}} \tag{5.75}$$

Using Eq. (5.75) the transversality condition of left optimisation interval end may be written as follows:

$$\psi_1(t_0^*) \cdot \delta x_1(t_0^*) + \psi_2(t_0^*) \cdot \delta x_2(t_0^*) + \bullet \bullet \\ + \psi_m(t_0^*) \cdot \delta x_m(t_0^*) = 0 \tag{5.76}$$

Then, we substitute (5.75) in (5.76)

$$\psi_1(t_0^*) \cdot \left\{ \begin{array}{l} -\dfrac{\left[\dfrac{\partial h_1(t_0^*)}{\partial x_2(t)}\right]_{t=t_0^*} \cdot \delta x_2(t_0^*)}{\left[\dfrac{\partial h_1(t_0^*)}{\partial x_1(t)}\right]_{t=t_0^*}} - \bullet \bullet + \\ -\dfrac{\left[\dfrac{\partial h_1(t_0^*)}{\partial x_m(t)}\right]_{t=t_0^*} \cdot \delta x_m(t_0^*)}{\left[\dfrac{\partial h_1(t_0^*)}{\partial x_1(t)}\right]_{t=t_0^*}} + \\ -\dfrac{\left[\dfrac{\partial h_1(t_0^*)}{\partial t}\right]_{t=t_0^*} \cdot \delta t_0^*}{\left[\dfrac{\partial h_1(t_0^*)}{\partial x_1(t)}\right]_{t=t_0^*}} \end{array} \right\}$$

$$+ \psi_2(t_0^*) \cdot \delta x_2(t_0^*) + \bullet \bullet + \psi_m(t_0^*) \cdot \delta x_m(t_0^*) = 0 \tag{5.77}$$

The further sequence is the transformation of (5.77) to a form, in which dependencies among conjugate variables are visible

$$\left[-\psi_1(t_0^*) \cdot \frac{\left[\dfrac{\partial h_1(t_0^*)}{\partial x_2(t)}\right]_{t=t_0^*}}{\left[\dfrac{\partial h_1(t_0^*)}{\partial x_1(t)}\right]_{t=t_0^*}} + \psi_2(t_0^*) \right] \cdot \delta x_2(t_0^*)$$

$$+ \left[-\psi_1(t_0^*) \cdot \frac{\left[\dfrac{\partial h_1(t_0^*)}{\partial x_3(t)}\right]_{t=t_0^*}}{\left[\dfrac{\partial h_1(t_0^*)}{\partial x_1(t)}\right]_{t=t_0^*}} + \psi_3(t_0^*) \right] \cdot \delta x_3(t_0^*)$$

$$+ \bullet \left[-\psi_1(t_0^*) \cdot \frac{\left[\dfrac{\partial h_1(t_0^*)}{\partial x_m(t)}\right]_{t=t_0^*}}{\left[\dfrac{\partial h_1(t_0^*)}{\partial x_1(t)}\right]_{t=t_0^*}} + \psi_m(t_0^*) \right] \cdot \delta x_m(t_0^*)$$

$$+ \left[-\psi_1(t_0^*) \cdot \frac{\left[\dfrac{\partial h_1(t_0^*)}{\partial t}\right]_{t=t_0^*} \cdot \delta t_0^*}{\left[\dfrac{\partial h_1(t_0^*)}{\partial x_1(t)}\right]_{t=t_0^*}} \right] = 0 \tag{5.78}$$

As differentials $\delta x_2(t_0^*), \delta x_3(t_0^*), \ldots, \delta x_m(t_0^*), \delta t_0^*$ may have different signs and $\psi_1(t_0^*) \neq 0$ taking into account (5.78) written out as elements $\psi_1(t_0^*) = C_1$, $\psi 2(t_0^*) = C_2, \ldots, \psi_m(t_0^*) = C_m$ it is possible to write the following relationships:

$$C_2 = C_1 \cdot \frac{\left[\frac{\partial h_1(t_0^*)}{\partial x_2(t)}\right]_{t=t_0^*}}{\left[\frac{\partial h_1(t_0^*)}{\partial x_1(t)}\right]_{t=t_0^*}} \quad C_3 = C_1 \cdot \frac{\left[\frac{\partial h_1(t_0^*)}{\partial x_3(t)}\right]_{t=t_0^*}}{\left[\frac{\partial h_1(t_0^*)}{\partial x_1(t)}\right]_{t=t_0^*}}$$

$$\bullet \; \bullet \quad C_m = C_1 \cdot \frac{\left[\frac{\partial h_1(t_0^*)}{\partial x_m(t)}\right]_{t=t_0^*}}{\left[\frac{\partial h_1(t_0^*)}{\partial x_1(t)}\right]_{t=t_0^*}} \tag{5.79}$$

$$C_1 \cdot \frac{\left[\frac{\partial h_1(t_0^*)}{\partial t}\right]_{t=t_0^*}}{\left[\frac{\partial h_1(t_0^*)}{\partial x_1(t)}\right]_{t=t_0^*}} = 0 \tag{5.80}$$

We write Eq. (5.80) in a new form

$$\boldsymbol{\psi}(t_0^*) = \boldsymbol{C} = \begin{bmatrix} 1 \\ \frac{\left[\frac{\partial h_1(t_0^*)}{\partial x_2(t)}\right]_{t=t_0^*}}{\left[\frac{\partial h_1(t_0^*)}{\partial x_1(t)}\right]_{t=t_0^*}} \\ \frac{\left[\frac{\partial h_1(t_0^*)}{\partial x_3(t)}\right]_{t=t_0^*}}{\left[\frac{\partial h_1(t_0^*)}{\partial x_1(t)}\right]_{t=t_0^*}} \\ \bullet \\ \frac{\left[\frac{\partial h_1(t_0^*)}{\partial x_m(t)}\right]_{t=t_0^*}}{\left[\frac{\partial h_1(t_0^*)}{\partial x_1(t)}\right]_{t=t_0^*}} \end{bmatrix} \cdot C_1 = \boldsymbol{G} \cdot C_1 \tag{5.81}$$

Now, Eqs. (5.41) and (5.42) receive the following form:

$$\hat{\boldsymbol{u}}(t - \boldsymbol{h}(t))_{(+)} = (\mathbf{A_1} + \mathbf{A_3})^{-1} \cdot \left[\mathbf{A}_1^{-1} \cdot \mathbf{B}(t) \cdot \mathbf{Y}(t) \cdot \mathbf{S} \cdot \boldsymbol{1} - \cdot \mathbf{S}_1{}^{\mathrm{T}} \cdot \boldsymbol{G} \cdot \hat{C}_1\right] \tag{5.82}$$

$$\hat{z}(t - \boldsymbol{\omega}) = \mathbf{A}_2^{-1} \cdot \mathbf{S}_2^{\mathrm{T}} \cdot \boldsymbol{G} \cdot \hat{C}_1 \tag{5.83}$$

Using Eq. (5.43), and (5.82) and (5.83), we may write the equation of the system reservoir state trajectories in the following form:

$$x(t) = x_0(t_0^*) + \int_{t_0^*}^{t} \begin{bmatrix} Q^P(\xi - \tau(\xi)) - \\ +\mathbf{S}_1 \cdot \hat{u}((\xi) - h(\xi)) + \\ +\mathbf{S}_2 \cdot \hat{z}(\xi - \omega) \end{bmatrix} d\xi \tag{5.84}$$

Further, we will aim to determine constant C_1. In Eq. (5.84) we substitute (5.82) and (5.83) and assume an upper integration boundary equal to unknown end optimisation time.

$$x(W) = x_0(t_0^*) + \int_{t_0^*}^{W} \begin{bmatrix} Q^P(t - \tau(t)) - \\ +\mathbf{S}_1 (\mathbf{A}_1 + \mathbf{A}_3)^{-1} \cdot \\ \cdot [\, \mathbf{A}_1^{-1} \cdot \mathbf{B}(t) \cdot \mathbf{Y}(t) \cdot \mathbf{S} \cdot \mathit{1} + \\ - \cdot \mathbf{S}_1^{\mathrm{T}} \cdot G \cdot \hat{C}_1 \,] + \\ +\mathbf{S}_2 \cdot \mathbf{A}_2^{-1} \cdot \mathbf{S}_2^{\mathrm{T}} \cdot G \cdot \hat{C}_1 \end{bmatrix} dt \tag{5.85}$$

We rearrange Eq. (5.85) integrating expressions constant in time

$$\begin{aligned} x(W) = & x_0(t_0^*) + \\ & + \int_{t_0^*}^{W} \begin{bmatrix} Q^P(t - \tau(t)) - \\ +\mathbf{S}_1 (\mathbf{A}_1 + \mathbf{A}_3)^{-1} \cdot \\ \cdot [\, \mathbf{A}_1^{-1} \cdot \mathbf{B}(t) \cdot \mathbf{Y}(t) \cdot \mathbf{S} \cdot \mathit{1} \end{bmatrix} dt + \\ & + [\, \mathbf{S}_1 (\mathbf{A}_1 + \mathbf{A}_3)^{-1} \cdot \mathbf{S}_1^{\mathrm{T}} + \mathbf{S}_2 \cdot \mathbf{A}_2^{-1} \cdot \mathbf{S}_2^{\mathrm{T}} \,] \\ & \cdot G \cdot \hat{C}_1 \cdot (\hat{W} - t_0^*) \end{aligned} \tag{5.86}$$

Then, we substitute (5.78) in the equation of bounded initial conditions (4.1)

$$\left[d_1(t_0^*) \bullet d_m(t_0^*) \right] \cdot \begin{bmatrix} x_1(t_0^*) \\ \bullet \\ x_m(t_0^*) \end{bmatrix} - b_1(t_0^*) = 0 \tag{5.87}$$

$$\begin{aligned} D(t_0^*) \cdot & \left\{ \begin{array}{l} x(W) - \\ \int_{t_0^*}^{W} \left\{ \begin{array}{l} Q^P(t - \tau(t)) - \\ + \mathbf{S}_1 (\mathbf{A}_1 + \mathbf{A}_3)^{-1} \cdot \\ \cdot \mathbf{A}_1^{-1} \cdot \mathbf{B}(t) \cdot \mathbf{Y}(t) \cdot \mathbf{S} \cdot 1 \end{array} \right\} dt + \\ - \left\{ \begin{array}{l} \mathbf{S}_1 (\mathbf{A}_1 + \mathbf{A}_3)^{-1} \cdot \mathbf{S}_1^T + \\ +\mathbf{S}_2 \cdot \mathbf{A}_2 \cdot \mathbf{S}_2^T \end{array} \right\} \cdot \\ \cdot G \cdot \hat{C}_1 \cdot \left(\hat{W} - t_0^* \right) \end{array} \right\} \\ & - b_1(t_0^*) = 0 \end{aligned} \tag{5.88}$$

After further transformations, we receive first relation specifying C_1 needed to determine optimisation end time t_0^*

$$\hat{C}_1 = \left\{ \begin{array}{l} D(t_0^*) \cdot \left\{ \begin{array}{l} \mathbf{S}_1 (\mathbf{A}_1 + \mathbf{A}_3)^{-1} \cdot \mathbf{S}_1^{\mathrm{T}} + \\ +\mathbf{S}_2 \cdot \mathbf{A}_2 \cdot \mathbf{S}_2^{\mathrm{T}} \end{array} \right\} \cdot \\ \cdot G \cdot \left(W - t_0^*\right) \end{array} \right\}^{-1} \cdot$$

$$\left\{ \begin{array}{l} D(t_0^*) \cdot \\ \left[\begin{array}{l} x(W) - \int\limits_{t_0^*}^{W} \left\{ \begin{array}{l} Q^P(t - \tau(t)) - \\ + \mathbf{S}_1 (\mathbf{A}_1 + \mathbf{A}_3)^{-1} \cdot \\ \cdot \mathbf{A}_1^{-1} \cdot \mathbf{B}(t) \cdot \mathbf{Y}(t) \cdot \mathbf{S} \cdot 1 \end{array} \right\} dt + \\ \qquad + b_1(t_0^*) \end{array} \right] \end{array} \right\} \tag{5.89}$$

In order to simplify formula (5.89), we will make the following substitutions

$$P_1 = S_1 (A_1 + A_3)^{-1} \cdot S_1^{\mathrm{T}} + S_2 \cdot A_2 \cdot S_2^{\mathrm{T}} \tag{5.90}$$

$$\boldsymbol{P}_2 = \int\limits_{t_0^*}^{W} \left\{ \begin{array}{l} Q^P(t - \tau(t)) - \\ + \mathbf{S}_1 (\mathbf{A}_1 + \mathbf{A}_3)^{-1} \cdot \\ \cdot \mathbf{A}_1^{-1} \cdot \mathbf{B}(t) \cdot \mathbf{Y}(t) \cdot \mathbf{S} \cdot 1 \end{array} \right\} dt \tag{5.91}$$

After substituting (5.90) and (5.93) in (5.81) we receive simple relationship describing variable C_1

$$\begin{aligned} \hat{C}_1 = &\left\{ D(t_0^*) \cdot \mathbf{P}_1 \cdot G \cdot (W - t_0) \right\}^{-1} \\ &\cdot \left\{ b_1(t_0^*) - D(t_0^*) \cdot (x(W) - P_2) \right\} \end{aligned} \tag{5.92}$$

In order to compute time t_0^* for optimisation start, another equation is necessary, which we may obtain using the following relationship:

$$H(t_0^*) - \left(\frac{\partial K \left[x(t_0^*), t_0^* \right]}{\partial t_0^*} \right) = 0 \tag{5.93}$$

In the case under consideration index (5.11) does not contain the function of initial conditions $K\left[x(t_0^*), t_0^*\right]$, therefore Hamiltonian function (5.93) $(H(t_0^*) = 0$.

Further procedure is the same as in the case of free optimisation end time $\hat{W}$, which as a consequence leads to finding:

- optimisation start time t_0^*,
- $\hat{\boldsymbol{u}}(\boldsymbol{t})$, vector of controlled outflows from reservoirs,
- $\hat{\boldsymbol{z}}(\boldsymbol{t})$, controlled transfers among reservoirs,
- taking into account delays in control and transfers,
- quality coefficient.

Chapter 6
Steady Optimisation Time, ST

6.1 Steady Optimisation Time, ST

The currently analysed issue is a compilation of the two former ones (Figs. 6.1, 6.2 and 6.3). Initial conditions of reservoirs are linked by equation $g_1(t_0)$, and final conditions for trajectories of reservoirs have to satisfy equation $g_2(W)$ with the assumption that t_0 and W are preset values determining a specific optimisation horizon. For the purposes of numerical illustration, a simple water distribution system was taken, distributing water from three reservoirs to three consumers, with the assumption that each reservoir supplies one consumer (Fig. 6.1).

6.2 Quality Index

We will assume a quality index (Lagrange's problem) in the following form:

$$F = \int_{t_0}^{W} f_0(u(t),\ x(t),\ t)\ dt \tag{6.1}$$

In particular for a system of three reservoirs without water transfers among them:

$$F = 0,5 \cdot \int_{t_0}^{W} \Big\{ a_1 \cdot [y_1(t) - u_1(t)]^2_{(+)} + a_2 \cdot [y_2(t) - u_2(t)]^2_{(+)} + a_3 \cdot [y_3(t) - u_3(t)]^2_{(+)} \Big\}\, dt \tag{6.2}$$

The physical meaning of this index (6.2) comes down to a summary of penalties incurred due to the difference in time $[t_0,\ W]$ among the required functions of water

W. Z. Chmielowski, *Management of Complex Multi-reservoir Water Distribution Systems Using Advanced Control Theoretic Tools and Techniques*, SpringerBriefs in Computational Intelligence, DOI: 10.1007/978-3-319-00239-2_6, © The Author(s) 2013

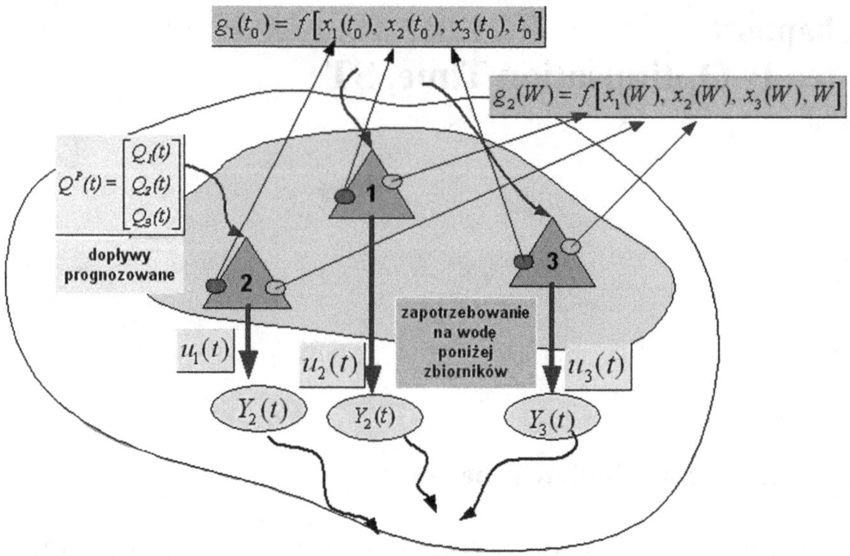

Fig. 6.1 The reservoir system

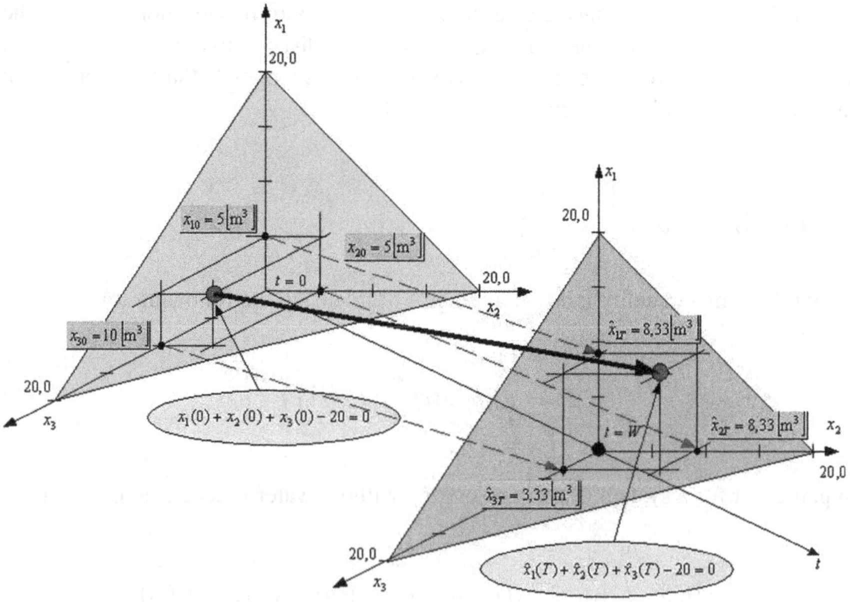

Fig. 6.2 Graphic illustration of the example

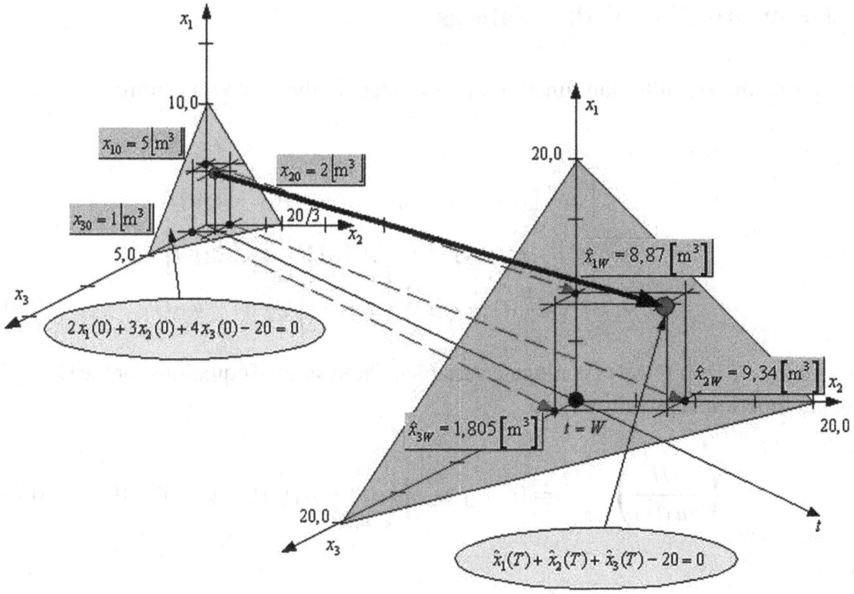

Fig. 6.3 Graphic illustration of the example

demand from reservoirs no. 1, 2, 3 $y_1(t)$, $y_2(t)$, $y_3(t)$ and effected controls (outflows from reservoirs) $\hat{u}_1(t)$, $\hat{u}_2(t)$, $u_3(t)$ $\forall t \in [t_0, W]$

6.3 The System State Equation

$$\begin{aligned} \dot{x}_1(t) &= Q_1^P(t) - u_1(t) \\ \dot{x}_2(t) &= Q_2^P(t) - u_2(t) \\ \dot{x}_3(t) &= Q_3^P(t) - u_3(t) \end{aligned} \tag{6.3}$$

The following have been assumed in Eq. (6.3):

- $\dot{x}_1(t)$, $\dot{x}_2(t)$, $\dot{x}_3(t)$—reservoir state derivative (change of water volume in the reservoir in time unit),
- $Q_1^P(t)$, $Q_2^P(t)$, $Q_3^P(t)$—predicted inflows to reservoirs,
- $u_1(t)$, $u_2(t)$, $u_3(t)$—outflows from reservoirs

Initial filling levels in reservoirs are linked by equation in form of e.g.:

$$g_1(t_0){:}d_1(t_0) \cdot x_1(t_0) + d_2(t_0) \cdot x_2(t_0) + d_3(t_0) \cdot x_3(t_0) - b_1(t_0) = 0 \tag{6.4}$$

Final filling levels in reservoirs are linked by equation in form of e.g.:

$$g_2(W){:}e_1(W) \cdot x_1(W) + e_2(W) \cdot x_2(W) + e_3(W) \cdot x_3(W) - b_2(W) = 0 \tag{6.5}$$

6.4 Optimisation Task Solutions

We create the Hamiltonian function of the system in the following form:

$$
\begin{aligned}
H &= -f_0 + \psi(t) \cdot f \\
H &= -0,5 \cdot \Big\{ a_1 \cdot [y_1(t) - u_1(t)]^2 + a_2 \cdot [y_2(t) - u_2(t)]^2 \\
&\quad + a_3 \cdot [y_3(t) - u_3(t)]^2 \Big\} + \psi_1(t) \cdot \left[Q_1^P(t) - u_1(t) \right] \\
&\quad + \psi_2(t) \cdot \left[Q_2^P(t) - u_2(t) \right] + \psi_3(t) \cdot \left[Q_3^P(t) - u_3(t) \right]
\end{aligned} \tag{6.6}
$$

where $\psi_1(t)$, $\psi_2(t)$, $\psi_3(t)$ conjugate variables The system of equations for the Hamiltonian function is as follows:

- $$\left(\frac{\partial H}{\partial u_1(t)} \right)_{\hat{u},\hat{x},\hat{\psi}} = 0 \qquad a_1 \cdot \left[y_1(t) - \hat{u}_1(t) \right] - \hat{\psi}_1(t) = 0 \tag{6.7}$$

- $$\hat{u}_1(t) = y_1(t) - \hat{\psi}_1(t) \big/ a_1 \tag{6.8}$$

- $$\left(\frac{\partial H}{\partial u_2(t)} \right)_{\hat{u},\hat{x},\hat{\psi}} = 0 \qquad a_2 \cdot \left[y_2(t) - \hat{u}_2(t) \right] - \hat{\psi}_2(t) = 0 \tag{6.9}$$

- $$\hat{u}_2(t) = y_2(t) - \hat{\psi}_2(t) \big/ a_2 \tag{6.10}$$

- $$\left(\frac{\partial H}{\partial u_3(t)} \right)_{\hat{u},\hat{x},\hat{\psi}} = 0 \qquad a_3 \cdot \left[y_3(t) - \hat{u}_3(t) \right] - \hat{\psi}_3(t) = 0 \tag{6.11}$$

- $$\hat{u}_3(t) = y_3(t) - \hat{\psi}_3(t) \big/ a_3 \tag{6.12}$$

- $$-\left(\frac{\partial H}{\partial x_1(t)} \right)_{\hat{u},\hat{x},\hat{\psi}} = \dot{\hat{\psi}}(t) \qquad \dot{\hat{\psi}}(t) = 0 \Rightarrow \hat{\psi}_1(t) = C_1 \tag{6.13}$$

- $$-\left(\frac{\partial H}{\partial x_2(t)} \right)_{\hat{u},\hat{x},\hat{\psi}} = \dot{\hat{\psi}}(t) \qquad \dot{\hat{\psi}}(t) = 0 \Rightarrow \hat{\psi}_2(t) = C_2 \tag{6.14}$$

- $$-\left(\frac{\partial H}{\partial x_3(t)} \right)_{\hat{u},\hat{x},\hat{\psi}} = \dot{\hat{\psi}}(t) \qquad \dot{\hat{\psi}}(t) = 0 \Rightarrow \hat{\psi}_3(t) = C_3 \tag{6.15}$$

- $$\left(\frac{\partial H}{\partial \psi_1(t)}\right)_{\hat{u},\hat{x}} = \dot{\hat{x}}_1(t) \qquad \dot{\hat{x}}_1(t) = Q_1^P(t) - \hat{u}_1(t) \tag{6.16}$$

- $$\left(\frac{\partial H}{\partial \psi_2(t)}\right)_{\hat{u},\hat{x}} = \dot{\hat{x}}_2(t) \qquad \dot{\hat{x}}_2(t) = Q_2^P(t) - \hat{u}_2(t) \tag{6.17}$$

- $$\left(\frac{\partial H}{\partial \psi_3(t)}\right)_{\hat{u},\hat{x}} = \dot{\hat{x}}_3(t) \qquad \dot{\hat{x}}_3(t) = Q_3^P(t) - \hat{u}_3(t) \tag{6.18}$$

For the boundary condition in form (6.4), acceptable shifts of initial conditions for trajectories of states have to satisfy the following condition:

$$\left(\frac{\partial g_1(t_0)}{\partial x_1(t_0)}\right) \cdot \delta x_1(t_0) + \left(\frac{\partial g_1(t_0)}{\partial x_2(t_0)}\right) \cdot \delta x_2(t_0) + \left(\frac{\partial g_1(t_0)}{\partial x_3(t_0)}\right) \cdot \delta x_3(t_0) = 0 \tag{6.19}$$

and thus:

$$\begin{aligned}
&\left(\frac{\partial\left(d_1(t_0)\cdot x_1(t_0) + d_2(t_0)\cdot x_2(t_0) + d_3(t_0)\cdot x_3(t_0) - b_1(t_0)\right)}{\partial x_1(t_0)}\right)\cdot \delta x_1(t_0) + \\
&+\left(\frac{\partial\left(d_1(t_0)\cdot x_1(t_0) + d_2(t_0)\cdot x_2(t_0) + d_3(t_0)\cdot x_3(t_0) - b_1(t_0)\right)}{\partial x_2(t_0)}\right)\cdot \delta x_2(t_0) + \\
&+\left(\frac{\partial\left(d_1(t_0)\cdot x_1(t_0) + d_2(t_0)\cdot x_2(t_0) + d_3(t_0)\cdot x_3(t_0) - b_1(t_0)\right)}{\partial x_3(t_0)}\right)\cdot \delta x_3(t_0) = 0
\end{aligned} \tag{6.20}$$

and

$$d_1(t_0)\cdot \delta x_1(t_0) + d_2(t_0)\cdot \delta x_2(t_0) + d_3(t_0)\cdot \delta x_3(t_0) = 0 \tag{6.21}$$

then, using Eq. (6.21) we obtain:

$$\delta x_1(t_0) = -\frac{d_2(t_0)}{d_1(t_0)}\cdot \delta x_2(t_0) - \frac{d_3(t_0)}{d_1(t_0)}\cdot \delta x_3(t_0) \tag{6.22}$$

Left transversality condition is satisfied if

$$\hat{\psi}_1(t_0)\cdot \delta x_1(t_0) + \hat{\psi}_2(t_0)\cdot \delta x_2(t_0) + \hat{\psi}_3(t_0)\cdot \delta x_3(t_0) = 0 \tag{6.23}$$

Substituting (6.22) in (6.23), we receive the following equation:

$$\left(\hat{\psi}_2(t_0) - \hat{\psi}_1(t_0)\cdot\frac{d_2(t_0)}{d_1(t_0)}\right)\cdot \delta x_2(t_0) = \left(\hat{\psi}_3(t_0) - \hat{\psi}_1(t_0)\cdot\frac{d_3(t_0)}{d_1(t_0)}\right)\cdot \delta x_3(t_0) \tag{6.24}$$

taking into account the fact that differentials $\delta x_2(t_0), \delta x_3(t_0)$ may have any values, we obtain:

$$\hat{\psi}_2(t_0) = \hat{\psi}_1(t_0) \cdot \frac{d_2(t_0)}{d_1(t_0)}, \qquad \hat{\psi}_3(t_0) = \hat{\psi}_1(t_0) \cdot \frac{d_3(t_0)}{d_1(t_0)} \tag{6.25}$$

By analogy, with reference to (6.5), permissible shifts of final conditions in the trajectories of states have to satisfy the following condition:

$$-\left(\frac{\partial g_2(W)}{\partial x_1(W)}\right) \cdot \delta x_1(W) - \left(\frac{\partial g_2(W)}{\partial x_2(W)}\right) \cdot \delta x_2(W) - \left(\frac{\partial g_2(W)}{\partial x_3(W)}\right) \cdot \delta x_3(W) = 0 \tag{6.26}$$

and thus:

$$\begin{aligned} &-\left(\frac{\partial(e_1(W)x_1(W)+e_2(W)x_2(W)+e_3(W)x_3(W)-b_2(W))}{\partial x_1(W)}\right) \cdot \delta x_1(W) + \\ &-\left(\frac{\partial(e_1(W)x_1(W)+e_2(W)x_2(W)+e_3(W)x_3(W)-b_2(W))}{\partial x_2(W)}\right) \cdot \delta x_2(W) + \\ &-\left(\frac{\partial(e_1(W)x_1(W)+e_2(W)x_2(W)+e_3(W)x_3(W)-b_2(W))}{\partial x_3(W)}\right) \cdot \delta x_3(W) = 0 \end{aligned} \tag{6.27}$$

$$-e_1(W) \cdot \delta x_1(W) - e_2(W) \cdot \delta x_2(W) - e_3(W) \cdot \delta x_3(W) = 0 \tag{6.28}$$

Using (6.28) we receive:

$$\delta x_1(W) = -\frac{e_2(W)}{e_1(W)} \cdot \delta x_2(W) - \frac{e_3(W)}{e_1(W)} \cdot \delta x_3(W) \tag{6.29}$$

Right transversality condition is satisfied if

$$-\hat{\psi}_1(W) \cdot \delta x_1(W) - \hat{\psi}_2(W) \cdot \delta x_2(W) - \hat{\psi}_3(W) \cdot \delta x_3(W) = 0 \tag{6.30}$$

Substituting (6.29) in (6.30), we receive the following equation:

$$\left(\hat{\psi}_1(W) \cdot \frac{e_2(W)}{e_1(W)} - \hat{\psi}_2(W)\right) \cdot \delta x_2(W) = \left(\hat{\psi}_3(W) - \hat{\psi}_1(W) \cdot \frac{e_3(W)}{e_1(W)}\right) \cdot \delta x_3(W) \tag{6.31}$$

taking into account the fact that differentials $\delta x_2(W), \delta x_3(W)$ may have any values, we obtain:

$$\hat{\psi}_2(W) = \hat{\psi}_1(W) \cdot \frac{e_2(W)}{e_1(W)}, \qquad \hat{\psi}_3(W) = \hat{\psi}_1(W) \cdot \frac{e_3(W)}{e_1(W)} \tag{6.32}$$

Equations (6.13), (6.14), (6.15) prove that:

$$\hat{\psi}_1(t) = C_1, \qquad \hat{\psi}_2(t) = C_2, \qquad \hat{\psi}_3(t) = C_3 \tag{6.33}$$

and using equations (6.25), (6.32), we obtain:

$$C_2 = C_1 \cdot \left(\frac{d_2(t_0)}{d_1(t_0)} + \frac{e_2(W)}{e_1(W)}\right)\Big/2, \quad C_3 = C_1 \cdot \left(\frac{d_3(t_0)}{d_1(t_0)} + \frac{e_3(W)}{e_1(W)}\right)\Big/2 \tag{6.34}$$

After integrating equations (6.16), (6.17), (6.18) we obtain the general form of the optimal state trajectories: (extremals), which $t = W$ and taking into account (6.34), get the following form:

$$\begin{aligned}
\hat{x}_1(W) &= \int_{t_0}^{W} \left[Q_1^P(\xi) - y_1(\xi)\right] d\xi + \tfrac{C_1 \cdot W}{a_1} + \hat{x}_1(t_0) \\
\hat{x}_2(W) &= \int_{t_0}^{W} \left[Q_2^P(\xi) - y_2(\xi)\right] d\xi + C_1 \cdot \tfrac{W}{a_2} \cdot \left(\tfrac{d_2(t_0)}{d_1(t_0)} + \tfrac{e_2(W)}{e_1(W)}\right)\Big/2 + \hat{x}_2(t_0) \\
\hat{x}_3(W) &= \int_{t_0}^{W} \left[Q_2^P(\xi) - y_2(\xi)\right] d\xi + C_1 \cdot \tfrac{W}{a_3} \cdot \left(\tfrac{d_3(t_0)}{d_1(t_0)} + \tfrac{e_3(W)}{e_1(W)}\right)\Big/2 + \hat{x}_3(t_0)
\end{aligned} \tag{6.35}$$

We substitute Eq. (6.35) in Eq. (6.5), and then subtract Eq. (6.4) from obtained equation, so as to consequently obtain the relationship determining constant C_1 in function of initial conditions

$$C_1 = \frac{\sum_{i=1}^{3} \left\{ \begin{array}{l} (d_i(t_0) - e_i(W)) \cdot x_i(t_0) \\ -e_i(W) \cdot \int_{t_0}^{W} \left[Q_1^P(t) - y_1(t)\right] dt \end{array} \right\} + (b_2(W) - b_1(t_0))}{W \cdot \left[\begin{array}{l} \frac{e_1(W)}{a_1} + \frac{e_2(W)}{a_2}\left(\frac{d_2(t_0)}{d_1(t_0)} + \frac{e_2(W)}{e_1(W)}\right)\Big/2+ \\ +\frac{e_3(W)}{a_3}\left(\frac{d_3(t_0)}{d_1(t_0)} + \frac{e_3(W)}{e_1(W)}\right)\Big/2 \end{array} \right]} \tag{6.36}$$

Using Eq. (6.36) and dependencies (6.34), we compute constants C_2, C_3 in function of any $x_1(t_0), x_2(t_0), x_3(t_0)$ satisfying Eq. (6.4). We substitute the computed constants C_1, C_2, C_3 in (6.7), (6.8) and (6.9) to obtain the formulas specifying optimal control.

$$\hat{u}_1(t) = y_1(t) - C_1/a_1, \qquad \hat{u}_2(t) = y_2(t) - C_2/a_2, \qquad \hat{u}_3(t) = y_3(t) - C_3/a_3 \tag{6.37}$$

Optimal reservoir state trajectories are described by the following equations:

$$\begin{aligned}
\hat{x}_1(t) &= \int_{t_0}^{t} \left\{ Q_1^P(\xi) - [y_1(\xi) - C_1/a_1] \right\} d\xi + \hat{x}_1(t_0) \\
\hat{x}_2(t) &= \int_{t_0}^{t} \left\{ Q_2^P(\xi) - [y_2(\xi) - C_2/a_2] \right\} d\xi + \hat{x}_2(t_0) \\
\hat{x}_3(t) &= \int_{t_0}^{t} \left\{ Q_3^P(\xi) - [y_3(\xi) - C_3/a_3] \right\} d\xi + \hat{x}_3(t_0)
\end{aligned} \tag{6.38}$$

We receive minimum quality coefficient value (6.2) for any $x_1(t_0), x_2(t_0), x_3(t_0)$ by way of substituting (6.37) in (6.2)

Example 1 If the planes of initial and final conditions are parallel, that is $d_1(t_0) = e_1(W), d_2(t_0) = e_2(W), d_3(t_0) = e_3(W)$, then the selection of starting points of trajectories $x_1(t_0), x_2(t_0), x_3(t_0)$, satisfying Eq. (6.4) is insignificant in the context of computing a constant, and formula (6.36) is simplified to the following form:

$$C_1 = \frac{\sum_{i=1}^{3} \left\{ -e_i(W) \cdot \int_{t_0}^{W} \left[Q_1^P(t) - y_1(t) \right] dt \right\} + (b_2(W) - b_1(t_0))}{T \cdot \left[\frac{e_1(W)}{a_1} + \frac{e_2(W)}{a_2} + \frac{e_3(W)}{a_3} \right]} \tag{6.39}$$

We will make the following assumptions:

- initial reservoir states linked by the equation
  $g_1(t_0) : x_1(t_0) + x_2(t_0) + x_3(t_0) - 20 = 0,\ d_1(t_0) = d_2(t_0) = d_3(t_0) = 1$
- final reservoir states linked by the equation
  $g_2(W) : x_1(W) + x_2(W) + x_3(W) - 20 = 0,\ e_1(W) = e_2(W) = e_3(W) = 1$
- optimisation horizon $t_0 = 0, \quad W = 10[\text{s}]$,
- inflows to reservoirs
  $Q_1^P(t) = 1,\ \ Q_2^P(t) = 2,\ \ Q_3^P(t) = 3[\text{m}^3/\text{s}], t \in [0,\ 10]$
- water demand below the reservoirs
  $y_1(t) = 2,\ \ y_2(t) = 3,\ \ y_3 = 4\ [\text{m}^3/\text{s}], t \in [0,\ 10]$
- weight coefficients $a_1 = 1, a_2 = 1, a_3 = 1$

For quality index in form of (6.2):
(a1) conjugate variable according to formula (6.39) is $\hat{\psi}_1(t) = C_1$

$$C_1 = \frac{\left\{ -\int_0^{10} [1-2]\ dt\ -\ \int_0^{10} [2-3]\ dt - \int_0^{10} [2-4]\ dt \right\} + (20-20)}{10 \cdot (1+1+1)} = 1,333$$

(a2) conjugate variables according to formula (6.34) are $\hat{\psi}_2(t) = C_2$, $\hat{\psi}_3(t) = C_3$

$$C_2 = C_3 = C_1 = 1,333$$

(a3) control according to formula (6.37)

$$\hat{u}_1(t) = 2 \cdot 1(t) - 1,333 = 0,666 \left[m^3 / s \right]$$
$$\hat{u}_2(t) = 3 \cdot 1(t) - 1,333 = 1,666 \left[m^3 / s \right]$$
$$\hat{u}_3(t) = 4 \cdot 1(t) - 1,333 = 2,666 \left[m^3 / s \right]$$

(a4) for any initial states $x_1(0), x_2(0), x_3(0)$ satisfying Eq. (6.4), the obtained final states of reservoirs will satisfy Eq. (6.5), at same (minimal) quality coefficient value, e.g. for $x_1(0) = 5, x_2(0) = 5, x_3 = (0) = 10$

$$\hat{x}_1(W) = 5 + \int_0^{10} (1 - 0,666)\, dt = 8,333 \left[m^3\right]$$
$$\hat{x}_2(W) = 5 + \int_0^{10} (2 - 1,666)\ dt = 8,333 \left[m^3\right]$$
$$\hat{x}_3(W) = 10 + \int_0^{10} (2 - 2,666)\ dt = 3,333 \left[m^3\right]$$

obtained end filling levels in reservoirs satisfy condition (6.5) $x_1(W) + x_2(W) + x_3(W) - 20 = 0$, and quality coefficient value is

$$F_{\min} = \int_0^{10} \left[(2 - 0,666)^2 + (3 - 1,666)^2 + (4 - 2,666)^2\right] dt = 45,59$$

(b1) or $x_1(0) = 1,\ \ x_2(0) = 4,\ \ x_3(0) = 15$

$$\hat{x}_1(W) = 1 + \int_0^{10} (1 - 0,666)\, dt = 4,333 \left[m^3\right]$$
$$\hat{x}_2(W) = 4 + \int_0^{10} (2 - 1,666)\ dt = 7,333 \left[m^3\right]$$
$$\hat{x}_3(W) = 15 + \int_0^{10} (2 - 2,666)\ dt = 8,333 \left[m^3\right]$$

obtained end filling levels in reservoirs also satisfy condition (6.5), for the same quality coefficient value

$$F_{\min} = \int_0^{10} \left[(2 - 0,666)^2 + (3 - 1,666)^2 + (4 - 2,666)^2\right] dt = 45,59$$

Example 2 If planes of initial and final conditions are not parallel, that is $d_1(t_0) \neq e_1(W),\ \ d_2(t_0) \neq e_2(W),\ \ d_3(t_0) \neq e_3(W)$, then the selection of starting points of trajectories $x_1(t_0), x_2(t_0), x_3(t_0)$ satisfying Eq. (6.4) is significant in computing constant C_1, according to formula (6.36)

We will make the following assumptions:

- initial reservoir states linked by the equation
  $g_1(t_0) : 2x_1(t_0) + 3x_2(t_0) + 4x_3(t_0) - 20 = 0,\ d_1(t_0) \neq d_2(t_0) \neq d_3(t_0) \neq 0$

- final reservoir states linked by the equation
$g_2(W): x_1(W) + x_2(W) + x_3(W) - 20 = 0,\; e_1(W) = e_2(W) = e_3(W) = 1$
- other input data unchanged.

(a1) for any initial states, currently satisfying equation $g_1(t_0)$ np. $x_1(0) = 2,\; x_2(0) = 2,\; x_3(0) = 2,5$
(a2) conjugate variable $\hat{\psi}(t) = C_1$ according to formula (6.36) has the following value:

$$C_1 = \frac{\left\{ \begin{array}{l} (2-1)\cdot 2 + (3-1)\cdot 2 + (4-1)\cdot 2,5 \\ -\int_0^{10}[1-2]\,dt - \int_0^{10}[2-3]\,dt - \int_0^{10}[2-4]\,dt \end{array} \right\} + (20-20)}{10\cdot\left(1 + 1\cdot\left(\frac{3}{2}+\frac{1}{1}\right)/2 + 1\cdot\left(\frac{4}{2}+\frac{1}{1}\right)/2\right)} = 1,427$$

(a3) conjugate variables are $\hat{\psi}_2(t) = C_2,\; \hat{\psi}_3(t) = C_3$

$$C_2 = C_1\cdot\left(\frac{d_2}{d_2}+\frac{e_2}{e_1}\right)\Big/2 = 1,427\cdot\left(\frac{3}{2}+\frac{1}{1}\right)/2 = 1,7834$$
$$C_3 = C_1\cdot\left(\frac{d_3}{d_2}+\frac{e_3}{e_1}\right)\Big/2 = 1,427\cdot\left(\frac{4}{2}+\frac{1}{1}\right)/2 = 2,141$$

(a4) control according to formula (6.37)

$$\hat{u}_1(t) = 2\cdot 1(t) - 1,427 = 0,573\cdot 1(t)\left[m^3/s\right],$$
$$\hat{u}_2(t) = 3\cdot 1(t) - 1,783 = 1,216\cdot 1(t)\left[m^3/s\right]$$
$$\hat{u}_3(t) = 4\cdot 1(t) - 2,141 = 1,859\cdot 1(t)\left[m^3/s\right]$$

(a5) and for assumed initial states, final states are as follows:

$$\hat{x}_1(10) = 2,0 + \int_0^{10}(1-0,573)\,dt = 6,27\left[m^3\right]$$
$$\hat{x}_2(10) = 2,0 + \int_0^{10}(2-1,216)\,dt = 9,83\left[m^3\right]$$
$$\hat{x}_3(10) = 2,5 + \int_0^{10}(2-1,859)\,dt = 3,91\left[m^3\right]$$

obtained end filling levels in reservoirs satisfy condition (6.5) $x_1(W) + x_2(W) + x_3(W) - 20 = 0$, and quality coefficient value is:

$$F_{\min} = \int_0^{10}\left[(2-0,573)^2 + (3-1,2166)^2 + (4-1,859)^2\right]dt = 98,03$$

(b1) for other initial states, currently satisfying equation $g_1(0)$ e.g.

$$x_1(0) = 5, \quad x_2(0) = 2, \quad x_3(0) = 1$$

(b2) conjugate variable $\hat{\psi}(t) = C_1$ is

$$C_1 = \frac{\left\{ \begin{array}{l} (2-1)\cdot 5 + (3-1)\cdot 2 + (4-1)\cdot 1 + \\ -\int_0^{10} [1-2]\ dt - \int_0^{10} [2-3]\ dt - \int_0^{10} [2-4]\ dt \end{array} \right\} + (20-20)}{10 \cdot \left(1 + 1 \cdot \left(\frac{3}{2} + \frac{1}{1}\right)/2 + 1 \cdot \left(\frac{4}{2} + \frac{1}{1}\right)/2\right)} = 1,387$$

(b3) conjugate variables $\hat{\psi}_2(t) = C_2$, $\hat{\psi}_3(t) = C_3$ are

$$C_2 = C_1 \cdot \left(\frac{d_2}{d_2} + \frac{e_2}{e_1}\right)\Big/ 2. = 1,387 \cdot \left(\frac{3}{2} + \frac{1}{1}\right)/2 = 1,734$$
$$C_3 = C_1 \cdot \left(\frac{d_3}{d_2} + \frac{e_3}{e_1}\right)\Big/ 2 = 1,387 \cdot \left(\frac{4}{2} + \frac{1}{1}\right)/2 = 2,0805$$

(b4) control according to formula (6.37)

$$\hat{u}_1(t) = 2 \cdot 1(t) - 1,387 = 0,613 \left[m^3 \big/ s\right]$$
$$\hat{u}_2(t) = 3 \cdot 1(t) - 1,734 = 1,266 \left[m^3 \big/ s\right]$$
$$\hat{u}_3(t) = 4 \cdot 1(t) - 2,0805 = 1,9195 \left[m^3 \big/ s\right]$$

(b5) and for assumed initial states, final states are as follows:

$$\hat{x}_1(10) = 5 + \int_0^{10} (1 - 0,613)\ dt = 8,87 \left[m^3\right]$$
$$\hat{x}_2(10) = 2 + \int_0^{10} (2 - 1,266)\ dt = 9,34 \left[m^3\right]$$
$$\hat{x}_3(10) = 1 + \int_0^{10} (2 - 1,9195)\ dt = 1,805 \left[m^3\right]$$

the obtained end filling levels in reservoirs satisfy condition (6.5) $x_1(W) + x_2(W) + x_3(W) - 20 = 0$, and quality coefficient value is:

$$F_{\min} = \int_0^{10} \left[(2 - 0,613)^2 + (3 - 1,266)^2 + (4 - 1,9195)^2\right] dt = 92,6$$

Chapter 7
Summary

We may draw the following conclusions as a result of many simulations carried out for systems varying in the structure of links, with reference both to reservoirs and conurbations, and to transfers among reservoirs, and various sets of delays concerning inflows, outflows and transfers:

1. The option to include and apply transfers among reservoirs considerably affects the operation of the combined reservoirs, primarily in the aspect of leaving the final states of the reservoirs at the required levels, and these states will be reached for minimal quality coefficient value. The option to take into account delays due to water transfers among reservoirs significantly improves the factual aspect of the applied solution.
2. Cooperation within a system of reservoirs without transfers among reservoirs comes down to operation of the reservoirs for which the shared purpose is to meet the water demands imposed on the system. None of the reservoirs supplying its share of the system water demands *sees* other reservoirs in the system. In some cases, cooperation of this sort may lead to a situation where, within a system of cooperating reservoirs, for unfavourable, low predicted inflow and after optimisation period *W*, some of reservoirs will remain with very low final states. This unfavourable effect may be alleviated precisely as a result of taking into account transfers among reservoirs, which, according to the optimisation task conditions, (value, transfer direction) will be selected so as to ensure required final states of the system reservoirs for a given vector of predicted inflows to the system of reservoirs.

Inclusion of inter-reservoir deployments within the structure of the system is crucial for:

- proper management of the system's water resources in the context of the equalisation of the final state of the reservoirs,
- even distribution of load between reservoirs in the system, resulting from the need to implement the function (vector, matrix) of demand,

W. Z. Chmielowski, *Management of Complex Multi-reservoir Water Distribution Systems Using Advanced Control Theoretic Tools and Techniques*, SpringerBriefs in Computational Intelligence, DOI: 10.1007/978-3-319-00239-2_7,

- additional water supplies for the indicated reservoir system, selected from the reservoirs belonging to the remainder of the system.

Supply is provided by the dispatcher's imposition of the structure of inter-reservoir deployments. By testing different structures of deployments, with identical input data into the system, the dispatcher is including the ability to determine the extent of changes in the final fullness of the reservoirs under certain demands on the water distribution system and the minimum value for the quality index. Having and using all the options resulting from the solutions obtained, the dispatcher can determine the optimal system state reservoir trajectories, the optimal trajectories of outflows from reservoirs and the optimal inter-reservoir trajectories of inter-reservoir deployments in the accepted optimization horizon, based on the standard data input to the system, such as:

- vector of inflows forecast for a specific time period,
- vector of initial fullness of the system reservoirs,
- the matrix of the function of demand,
- the vector of reservoir states required in the optimization horizon,

and by determining the value of the weighting factors, and the limitations of the control structure of the system including inter-reservoir deployments.

The optimal control problem presented in subsequent chapters are solved by analysis. The feature vector in the input parameters shown in the optimal controls in the system, the vector of optimal states of the system, and the optimal vector for inter-reservoir deployments. This solution is based on the clear and understandable nature of controlling outflows from reservoirs in the system of any given structure of relationships between its elements. The book presents the solution for a number of optimization tasks, based on which control algorithms have been developed for outflows from reservoirs for complex water management systems. The task optimizations presented (indicating optimal solutions adopted in the light of the performance index and constraints), may be included in optimization modules, which is one of the elements of the model for automation control of decision-making. Decisions on the system reservoir outflows can be taken with a full interaction by the dispatcher in the necessary moments of operation in the model or optionally in special situations from his point of view. Optimization tasks and the algorithms based on them cover the short-term planning layer of retention in the reservoir system under normal operation. The paper presents a class of optimization problems, particularly dynamic programming tasks with a square indicator of quality. The maximum continuous principle is used to solve the issue using vector-matrix algebra.

The approach presented for controlling complex multi-reservoir water systems may be included among the classical modelling of control methods that are used today. This approach places great emphasis on the accuracy of reflecting the phenomenon, which is causing problems with describing and modelling of complex processes in our environment. In the year 1965, in the scientific journal "Information and Control", the American computer scientist Lofti A. Zadeh published an article entitled "Fuzzy Sets" representing the assumptions of fuzzy set theories. Then

Zadeh developed and presented a complete theory of fuzzy sets. The mathematical formulation introduced by Zadeh allowed for the formulation and creation of fuzzy systems. The operation of these systems is based on explicit knowledge of the problem based on fuzzy rules in the form of conditional if-then statements. The premises and conclusions of the rules are often defined by fuzzy sets. The theory presented substantially expanded classical set theory, which assumed that an element can either belong or not belong to a set. According to the theory of fuzzy sets (basically) the item can belong to the set to "some extent", set at the value of the membership function. The proposed theory thus replaces zero—one logic (1= element belongs to the set, 0= element not in the set), to the "soft" logic 'an element belongs to a set with a membership function value in the range [0,1]. Ten years after Zadeha's first publication, Ebrahim Mamdani proposed an application of fuzzy sets in control technologies. In the control system he introduced a regulator acting on the basis of fuzzy logic with cause-effect relationships for the input and output signals of the controller: IF ()' THEN (…), e.g. "IF the inflow is too small THEN the outflow will be greater". Subsequently, over the years the concepts of regulators and their architecture were refined. This is how the Tagake-Sugeno architecture came to be commonly used in control systems. Since the first studies in this field have passed from the top of 45 years and is now the literature on the theory of fuzzy sets and their applications in various fields of industry, medicine and daily life is estimated at more than 20,000 items. Over the years, concepts have developed of the already popular fuzzy systems, fuzzy controllers and control systems using fuzzy controllers operating on the basis of fuzzy logic. Fuzzy systems consist of techniques and methods that are used for processing information received from the system's environment that is imprecise, vague or unspecific. Fuzzy systems allow description of a phenomenon as a wild card, which two-valued logic (binary) cannot recognize. Fuzzy systems are characterized in that the knowledge provided in the system is converted into symbolic form and stored in the databases of the rules using if-then conditions. Fuzzy systems are used everywhere there we do not have sufficient knowledge of the mathematical model governing a given phenomenon and where recreation of that model is not fully or completely possible. Processes occurring in the control complex multi-reservoir water systems certainly cannot be entirely accurately described by rigorous mathematical associations. In particular, analytical solutions are characterized by a large number of approximations and simplifications in relation to reality. It would appear that in the distant future the application in this field of classical control methods enriched with fuzzy logic algorithms will significantly improve the effectiveness of control of complex inter-reservoir water systems.

References

1. W. Chmielowski, Optymalizacja pracy systemu zbiornikow retencyjnych przy warunku koniecznosci zaspokojenia potrzeb. Czasopismo Techniczne Politechniki Krakowskiej, (Zeszyt î-8/2004), 2004
2. W. Chmielowski, Czas obserwacji i rozmaitoscwarunkow brzegowych na trajektoriach stanow, jako elementy optymalnego sterowania odplywami z systemu zbiornikow retencyjnych. Czasopismo Techniczne Politechniki Krakowskiej, (Zeszyt î-8/2005) (2005)
3. W. Chmielowski, Ograniczenia stanu i sterowania jako element optymalizacji wielozbiornikowego systemu wodno-gospodarczego (czesc i). Czasopismo Techniczne Politechniki Krakowskiej, (Zeszyt î-16/2005) (2005)
4. W. Chmielowski, Ograniczenia stanu i sterowania jako element optymalizacji wielozbiornikowego systemu wodno-gospodarczego (czesc ii). Czasopismo Techniczne Politechniki Krakowskiej, (Zeszyt î-16/2005) (2005)
5. W. Chmielowski, Zastosowania optymalizacji w gospodarce wodnej. Wydawnictwo Techniczne Politechniki Krakowskiej (2005)
6. W. Chmielowski, Model optymalizacyjny dystrybucji wody z systemu polaczonych zbiornikow retencyjnych. czesc i (ustalony horyzont optymalizacji). AMCS 2012 (2012)
7. W. Chmielowski, R. Uryga, Szczegolna postac warunkow brzegowych w sterowaniu optymalnym wielozbiornikowymi systemami wodnogospodarczymi. Czasopismo Techniczne Politechniki Krakowskiej, (Zeszyt î-1/2009) (2009)
8. W. Chmielowski, R. Uryga, Sterowanie systemem zbiornikow retencyjnych z uwzglednieniem przerzutow miedzyzbiornikowych przy okreslonych warunkach na trajektoriach stanow. czesc i. Czasopismo Techniczne Politechniki Krakowskiej, (Zeszyt î-2/2011) (2011)
9. W. Chmielowski, R. Uryga, Sterowanie systemem zbiornikow retencyjnych z uwzglednieniem przerzutow miedzyzbiornikowych przy okreslonych warunkach na trajektoriach stanow. czesc ii. Czasopismo Techniczne Politechniki Krakowskiej, (Zeszyt î-2/2011) (2011)
10. W. Findeisen, J. Szymanowski, A. Wierzbicki, Teoria i metody obliczeniowe, Warszawa PWN (1980)
11. H. Gorecki, Optymalizacja systemow dynamicznych, Warszawa PWN (1993)
12. H. Gorecki, Optymalizacja i sterowanie systemow dynamicznych. Wydawnictwo AGH (2006)
13. J. Polomska, W. Chmielowski, Optimization and control of water distribution in water management system characterized by specified quality and quantint requiremenst. Zeszyty naukowe Wydzialu elektroniki, telekomunikacji i informatyki Politechnik Gdanskiej, (Tom 1/2011) (2011)

W. Z. Chmielowski, *Management of Complex Multi-reservoir Water Distribution Systems Using Advanced Control Theoretic Tools and Techniques*, SpringerBriefs in Computational Intelligence, DOI: 10.1007/978-3-319-00239-2, © The Author(s) 2013